RETEACHING COPYMASTERS

Barbara L. Power

McDougal Littell

Evanston, Illinois • Boston • Dallas

International Standard Book Number: 0-395-87205-7
2 3 4 5 6 7 8 9 10 BEI 01 00 99 98 97

Reteach
Chapter 1

Name ____________________________________

What you should learn:

1.1	How to explore optical illusions and explore ways to use illusions

Correlation to Pupil's Textbook:

Mid-Chapter Partner Quiz (p. 29)
Exercises 1, 2

Chapter Checkup (p. 53)
Exercises 4–7

Examples *Exploring Optical Illusions and Using Illusions*

a. In the figures at the right, which solid dot is larger?

The solid dots are the same size. The solid dot on the right appears smaller because it is centered in a wide border.

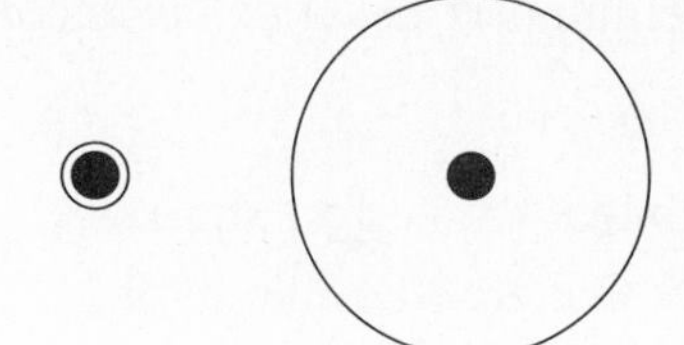

b. In the drawing shown at the right, compare the height of the top hat with the width of its brim.

The height of the top hat is the same as the width of the brim, but the eye will see a vertical line as longer than a horizontal line and an unbroken line as longer than a line intersecting with other lines.

Guidelines:

- Because optical illusions can fool your perception, you must train your eyes and mind to see reality.

EXERCISES

1. In the figures shown at the right, the figure A appears to be smaller than the figure B, although they are the same size. Why do you think the lower curve of figure A looks smaller than the lower curve of figure B?

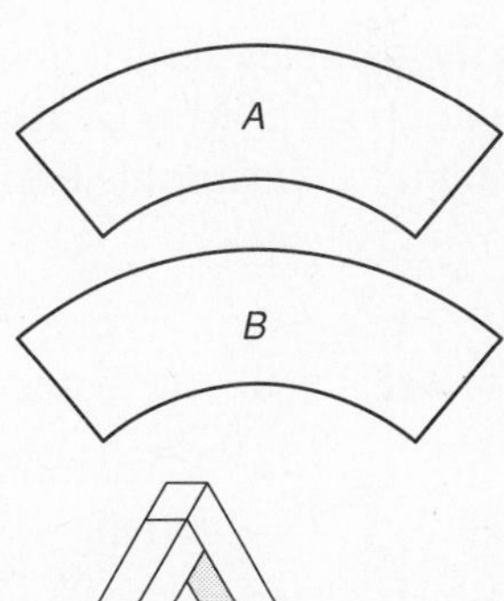

2. How would you describe the figure at the right when you first look at it? Now describe how the bars meet at each corner. Why is "impossible tri-bar" a good name for the figure?

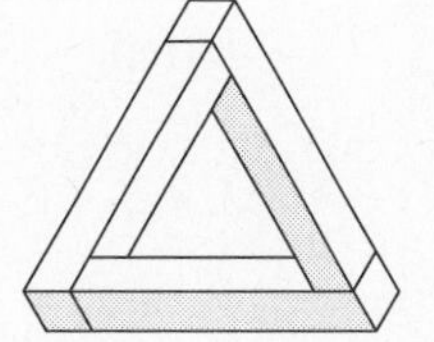

3. The impossible four-bar at the right can be drawn from a combination of bars from the two frames to its right. Which bars were used?

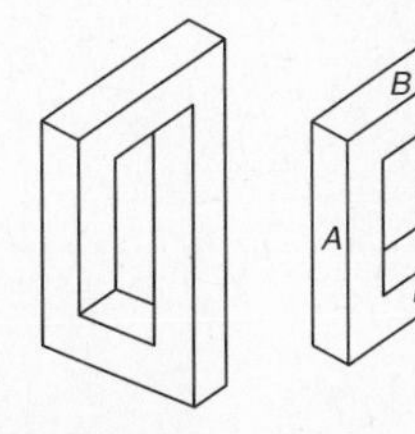
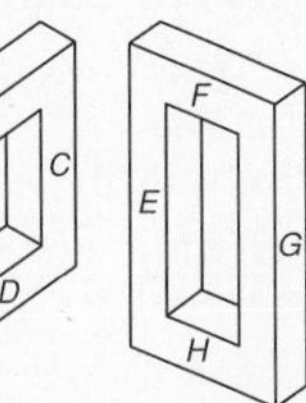

Reteach
Chapter 1

Name ______________________________

What you should learn:

1.2 How to sort a set of objects and identify attributes by examples

Correlation to Pupil's Textbook: Exercises None

Mid-Chapter Partner Quiz (p. 29)
Exercises 3–10

Examples *Sorting a Set of Objects and Identifying Attributes by Examples*

a. Classify the numbers listed below according to the following attributes.

$-13, 16, -8, 1, -3, 10, -5, 9$

Answers:

1. Numbers that are perfect squares — 1, 9, 16
2. Numbers that are divisible by 2 — 16, −8, 10
3. Numbers that are negative — −13, −8, −3, −5

b. Two figures are shown below as Type 1 and two figures are shown as Type 2. Which of the figures at the right are Type 1?

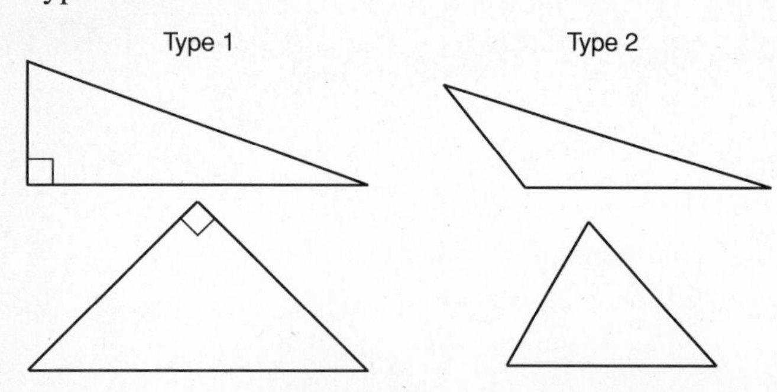

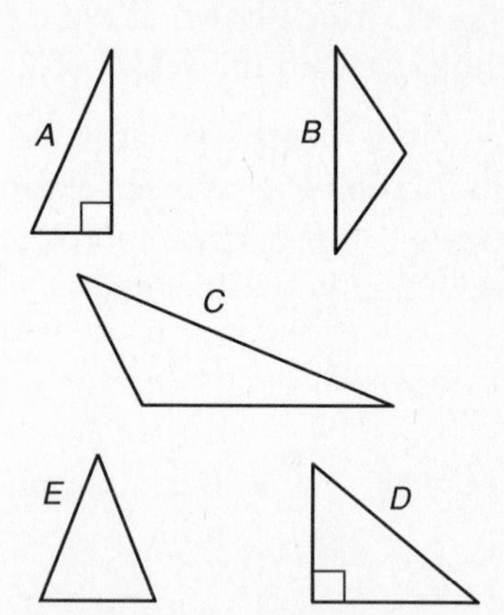

From the samples shown above, the distinguishing attribute about a Type 1 appears to be that the triangle has a right angle. Using this attribute, A and D are Type 1.

Guidelines:

- Attributes can be used to sort a set of objects.
- A Venn diagram can be used to represent the way a set is sorted.

EXERCISE

Classify the numbers in Example a above according to the following attributes.

1. Numbers that are odd
2. Numbers that are positive

Reteach
Chapter 1

Name ______________________________

What you should learn:

1.3	How to explore the concepts of congruence and similarity

Correlation to Pupil's Textbook:

Mid-Chapter Partner Quiz (p. 29)
Exercises 11, 12, 16, 17

Chapter Checkup (p. 53)
Exercises 1, 2, 4, 5

Examples *Identifying Congruent Figures and Similar Figures*

a. Divide the region at the right into two congruent parts. Two figures are congruent if they have the same shape and the same size.

Several solutions to this problem are shown below.

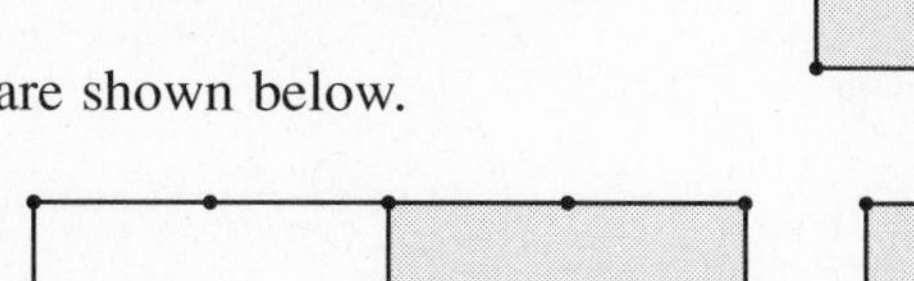

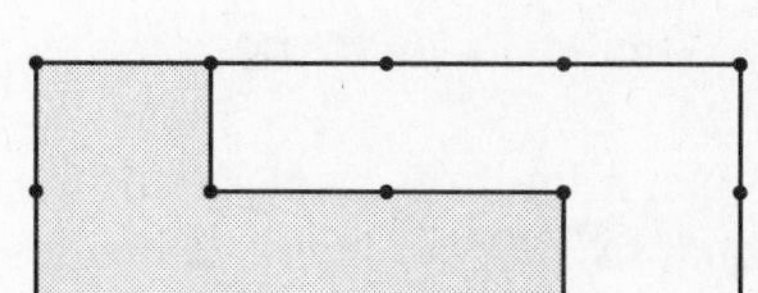

b. You are copying a newspaper article on a photocopier that enlarges an original to 121% enlargement. Is the enlargement congruent to the original or similar to the original? Explain your reasoning.

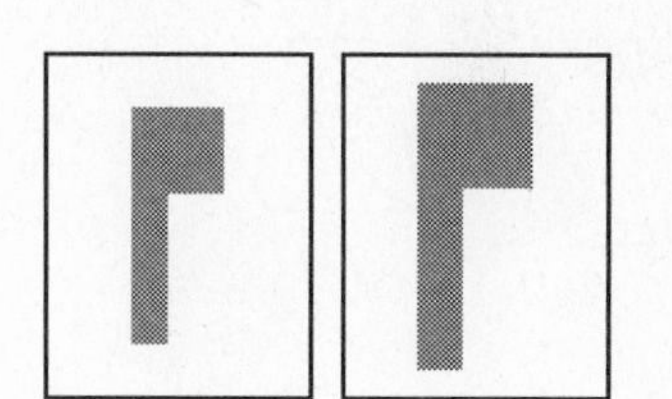

The enlargement is similar to the original because it has the same shape as the original but is larger. The enlargement is not congruent to the original because congruent figures must have the same shape and the same size.

Guidelines:

- If two figures are *congruent*, then either can be moved, turned, or flipped so that it coincides with the other figure.

EXERCISES

1. Find other solutions for Example a above.

2. In the figure at the right, two of the shapes are similar. Which two are similar?

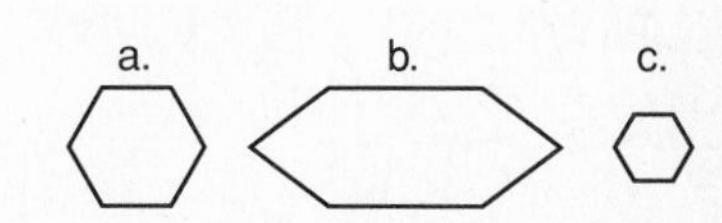

3. In the figure at the right, two of the shapes are congruent. Which two are congruent?

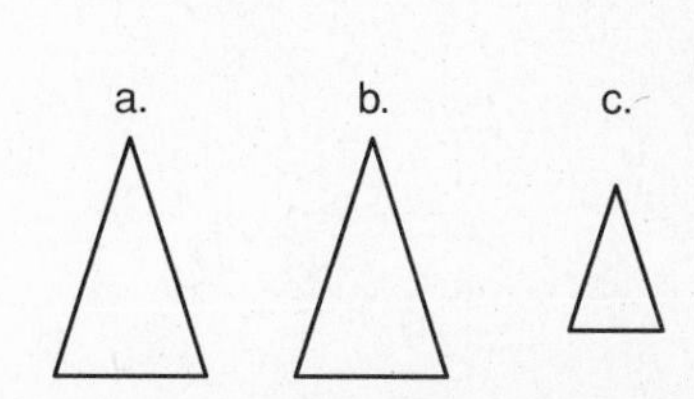

Reteach
Chapter 1

Name ______________________

What you should learn:

1.4	How to identify line symmetry and rotational symmetry in a figure

Correlation to Pupil's Textbook:

Mid-Chapter Partner Quiz (p. 29)	Chapter Checkup (p. 53)
Exercises 3–10, 13, 14	Exercises 6, 7

Examples *Identify Symmetric Objects*

a. Match the figures below with the correct lines of symmetry.

1. 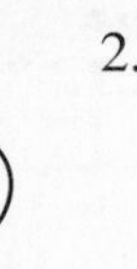
2. 
3. 

A. Vertical line of symmetry
B. Horizontal line of symmetry
C. No line of symmetry

Answers: 1. A, 2. C, 3. A, B

b. When a figure is rotated through a positive angle that is less than or equal to 180° so that it coincides with itself, it is said to have rotational symmetry. Identify the figures below that have rotational symmetry.

1. 

The figure has rotational symmetry. It will coincide with itself after being rotated 120°.

2. 

The figure has rotational symmetry. It will coincide with itself after being rotated 90° and 180°.

Guidelines:

- A figure has *line symmetry* or *reflection symmetry* if it can be divided into two parts, each of which is the mirror image of the other.
- A figure has *rotational symmetry* if it coincides with itself after rotating through a positive angle that is less than or equal to 180°.

EXERCISES

Identify any symmetry of the figures.

1. 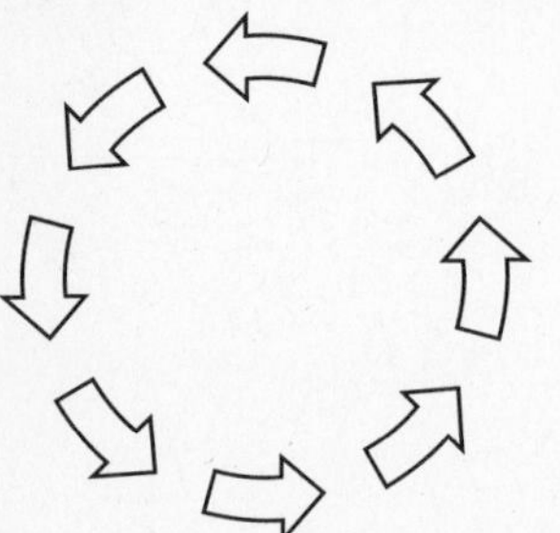

2. 

3. 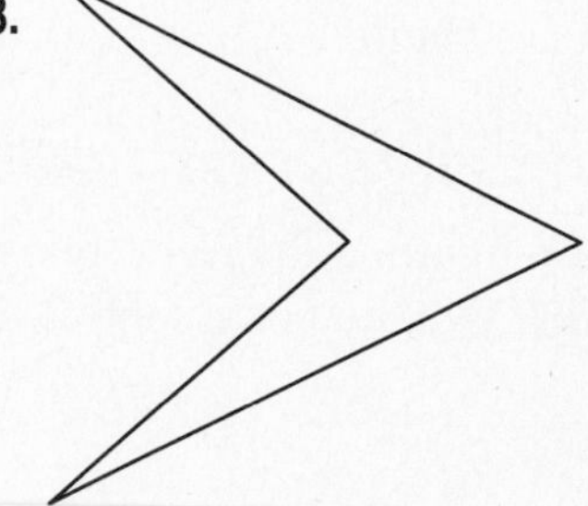

Reteach
Chapter 1

Name ____________________

What you should learn:

1.4 How to find the midpoint of a line segment in coordinate geometry

Correlation to Pupil's Textbook:

Mid-Chapter Partner Quiz (p. 29)	Chapter Checkup (p. 53)
Exercise 15	Exercise 3

Examples *Finding the Midpoint of a Line Segment*

a. Find the midpoint of the longest side of the triangle at the right.

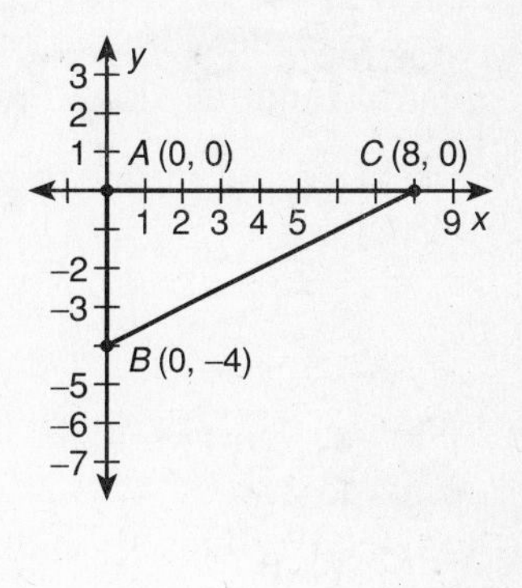

The midpoint of $\overline{BC}$, the longest side of triangle ABC, can be found using the Midpoint Formula.

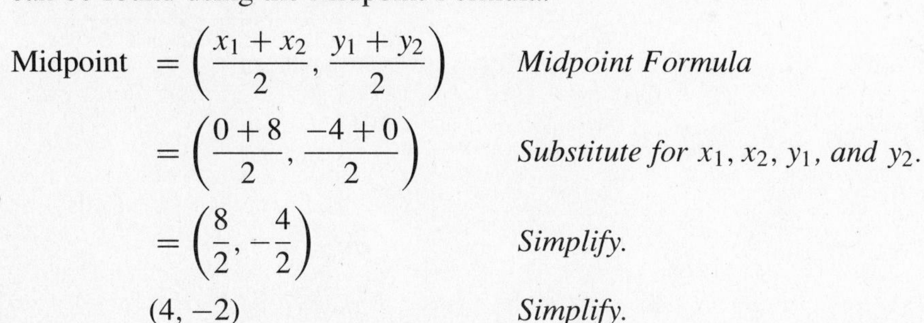

Midpoint $= \left(\dfrac{x_1 + x_2}{2}, \dfrac{y_1 + y_2}{2}\right)$ *Midpoint Formula*

$= \left(\dfrac{0+8}{2}, \dfrac{-4+0}{2}\right)$ *Substitute for x_1, x_2, y_1, and y_2.*

$= \left(\dfrac{8}{2}, -\dfrac{4}{2}\right)$ *Simplify.*

$(4, -2)$ *Simplify.*

b. Use the Midpoint Formula to find the midpoint of line segment $\overline{DE}$ where D is the point $(-13, 3)$ and E is the point $(-2, -4)$.

Midpoint $= \left(\dfrac{x_1 + x_2}{2}, \dfrac{y_1 + y_2}{2}\right)$ *Midpoint Formula*

$= \left(\dfrac{-13 + (-2)}{2}, \dfrac{3 + (-4)}{2}\right)$ *Substitute for x_1, x_2, y_1, and y_2.*

$= \left(-\dfrac{15}{2}, -\dfrac{1}{2}\right)$ *Simplify.*

Guidelines:

- The *midpoint* of a line segment bisects the segment into two shorter segments of equal length.

EXERCISES

1. Find the midpoint of each side of figure $ABCD$.

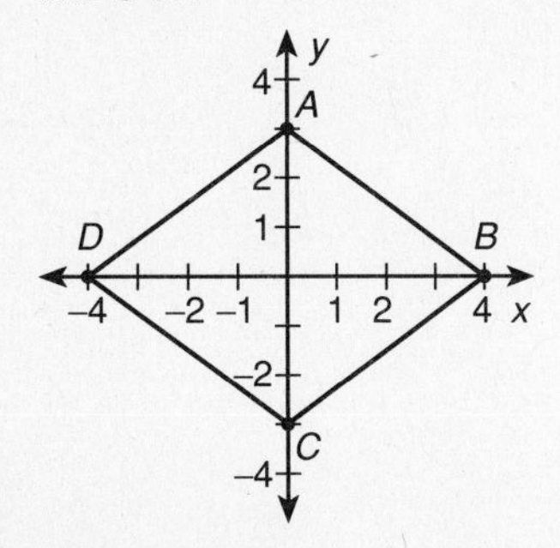

2. Let G be the midpoint of segment $\overline{FH}$. $\overline{GH}$ has a length of 7. What is the length of $\overline{FH}$?

Reteach
Chapter 1

Name ______________________________

What you should learn:

1.5 How to use slope in coordinate geometry

Correlation to Pupil's Textbook:

Chapter Checkup (p. 53)
Exercises 8, 9

Examples *Using Slope in Coordinate Geometry*

a. Line p passes through the points D and E, graphed in a coordinate plane. Find the slope of line p.

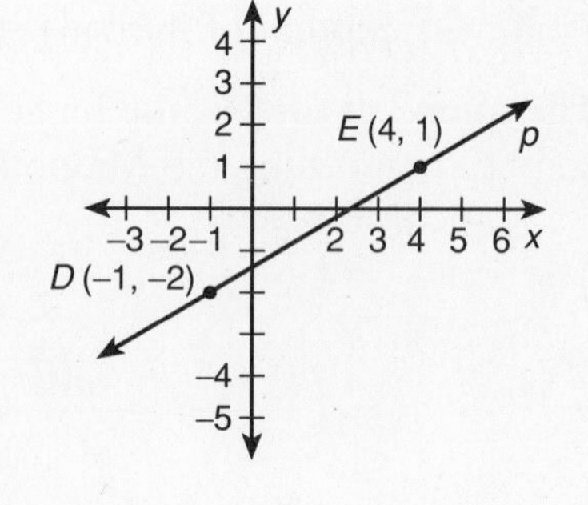

The slope, m, of line p can be found by using the formula for slope.

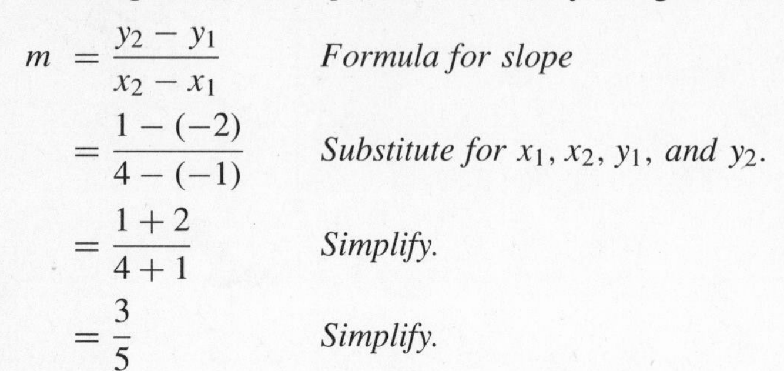

$$m = \frac{y_2 - y_1}{x_2 - x_1} \qquad \textit{Formula for slope}$$

$$= \frac{1-(-2)}{4-(-1)} \qquad \textit{Substitute for } x_1, x_2, y_1, \textit{ and } y_2.$$

$$= \frac{1+2}{4+1} \qquad \textit{Simplify.}$$

$$= \frac{3}{5} \qquad \textit{Simplify.}$$

The slope, m, of the line is $\frac{3}{5}$.

b. Find the slope of line q which is perpendicular to line p in Example a.

Two nonvertical lines with slopes of m_1 and m_2 are perpendicular if and only if $m_1 \cdot m_2 = -1$. Since the slope, m_1, of line p is $\frac{3}{5}$, then the slope, m_2, of line q is $-\frac{5}{3}$.

c. Find the slope of line t which is parallel to line p in Example a.

Two nonvertical lines are parallel if and only if they have the same slope. The slope of line t must be $\frac{3}{5}$ which is the slope of line p.

Guidelines:

- The *slopes* of perpendicular lines are negative reciprocals of each other.
- The line $y = mx + b$ has slope equal to m.

EXERCISES

1. Find the slope of the line h that passes through the points $A(-2, 5)$ and $B(3, 0)$.

2. Find the slope of a line perpendicular to line h in Exercise 1.

3. Find the slope of a line parallel to line h in Exercise 1.

4. Determine whether the lines are parallel, perpendicular, or neither.

a. Line p: $y = -x + 1$
Line q: $y = x + 1$

b. Line s: $y = \frac{1}{3}x - 2$
Line t: $y = -\frac{1}{3}x + 2$

c. Line m: $y = x + 3$
Line n: $y = x - 3$

Reteach
Chapter 1

Name ______________________________

What you should learn:

1.5	How to explore relationships in noncoordinate geometry

Correlation to Pupil's Textbook:

Chapter Checkup (p. 53)

Exercises 10, 11

Examples *Exploring Relationships in Noncoordinate Geometry*

a. It appears that planes P and R are perpendicular to plane Q. Match the following apparent relationships.

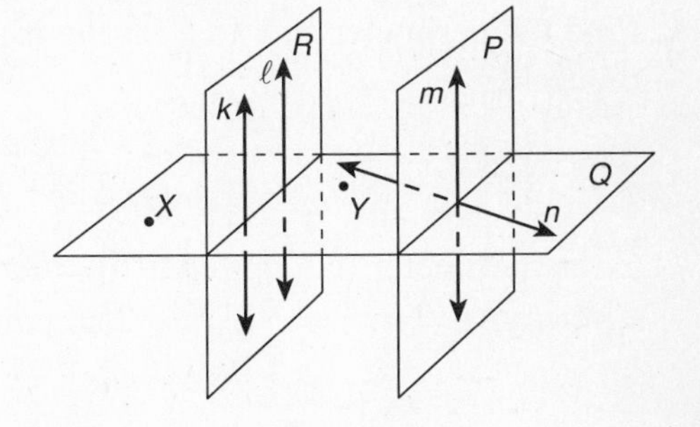

1. Line k and line ℓ	a. coplanar points
2. Line n and line m	b. perpendicular lines
3. Plane P and plane R	c. parallel lines
4. Point X and point Y	d. parallel planes

Answers: 1. c 2. b 3. d 4. a

b. Use the words perpendicular, parallel, and coplanar to describe the following relationships.

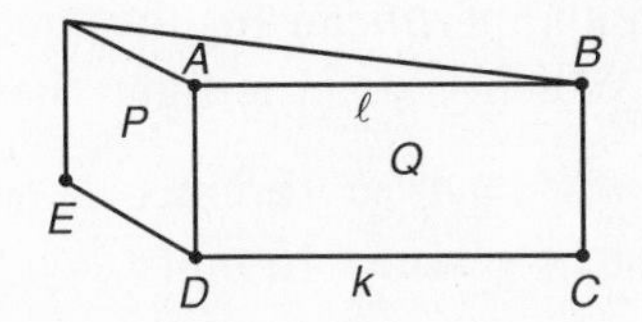

	Answers:
1. Points A, B, and D	coplanar
2. Plane P and line k	perpendicular
3. Lines ℓ and k	parallel
4. Planes P and Q	perpendicular
5. Points A, D, and E	coplanar

Guidelines:

- The type of geometry that was developed without a coordinate system is called *Euclidian geometry* or *noncoordinate geometry.*
- Points and/or lines that lie in one plane are said to be *coplanar.*
- Points are *collinear* if they lie on the same line.

EXERCISES

1. Draw two planes that are parallel, and draw a line in each plane so that the lines are not parallel to each other.
2. Draw two planes that are perpendicular, and draw a line in each plane so that the lines are parallel to each other.
3. Use the words perpendicular, parallel, and coplanar to describe the following relationships.

   a. Lines s and r

   b. Points B, C, and E

   c. Lines s and t

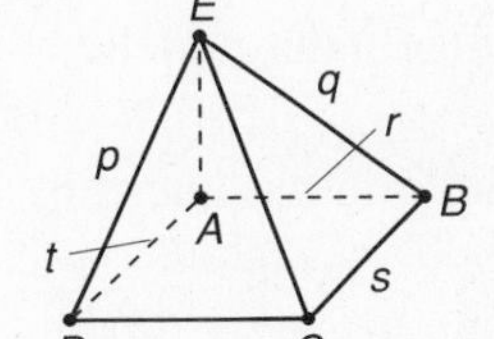

Reteach

Chapter 1

Name ______________________

What you should learn:

1.6	How to explore perimeter and area and use perimeter and area formulas in real-life

Correlation to Pupil's Textbook:

Chapter Checkup (p. 53)
Exercises 12–16

Examples *Exploring Perimeter and Area and Solving Real-life Problems*

a. Find the perimeter and area for the triangle and circle shown at the right.

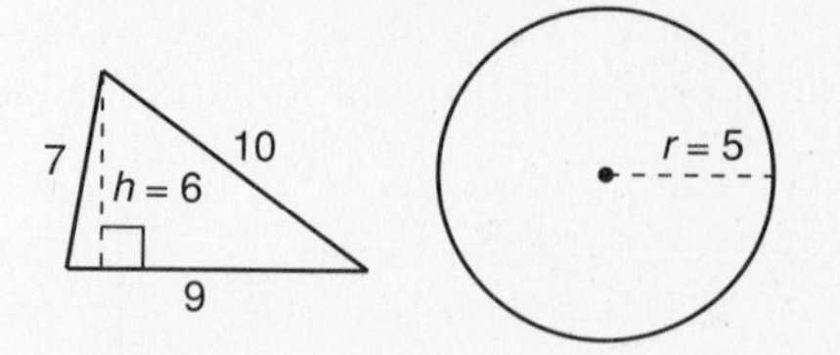

Triangle perimeter $= a + b + c = 7 + 9 + 10 = 26$ units
Triangle area $= \frac{1}{2}bh = \frac{1}{2}(9)(6) = 27$ units^2
Circle perimeter (circumference) $= 2\pi r = 2\pi(5) = 10\pi$ units
Circle area $= \pi r^2 = \pi(5)^2 = 25\pi$ units^2

b. Find the area and perimeter for the square and rectangle shown at the right. Which geometric figure has the greater perimeter?

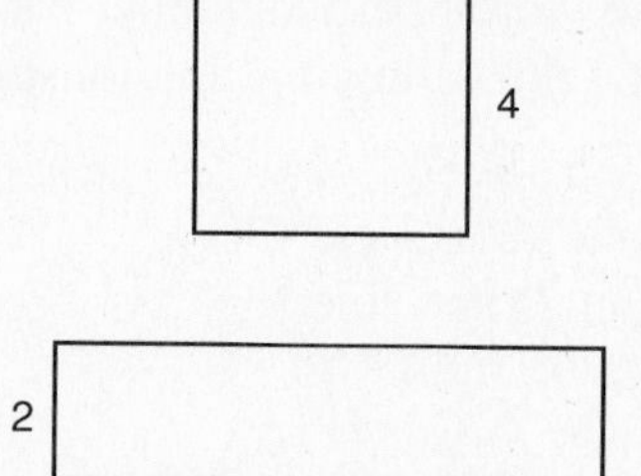

Square area $= s^2 = 4^2 = 16$ units^2
Square perimeter $= 4s = 16$ units

Rectangle area $= lw = (8)(2) = 16$ units^2
Rectangle perimeter $= 2l + 2w = 2(8) + 2(2)$
$= 16 + 4 = 20$ units

The rectangle has the greater perimeter.

Guidelines:

- *Perimeter* is a one-dimensional measure that uses units such as meters, inches, kilometers, and miles.
- *Area* is a two-dimensional measure that uses units such as square meters, square feet, and square miles.

EXERCISES

Use a problem-solving plan to find the number of square feet of linoleum needed to cover a floor that is 10.5 feet by 11.5 feet.

1. Write a verbal model.

2. Assign labels.

3. Write an algebraic model.

4. Solve the algebraic model.

5. Answer the original question and check your answers.

Reteach
Chapter 1

Name ____________________

What you should learn:

1.7	How to use a straightedge and compass to copy and bisect a segment or angle

Examples — *Copying and Bisecting Segments and Angles*

a. Copy segment $\overline{BC}$, then bisect it.

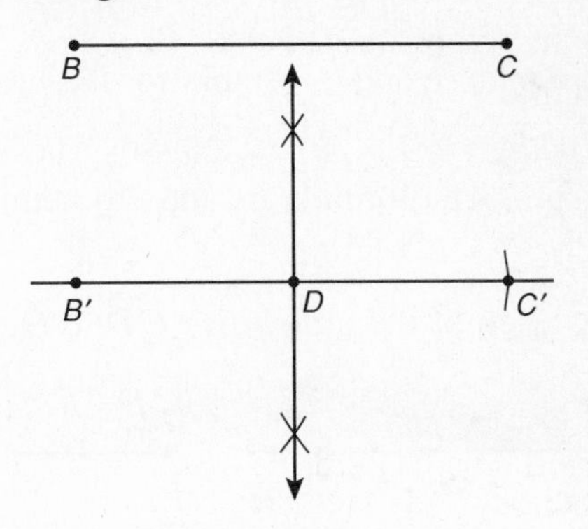

Draw a segment longer than $\overline{BC}$ and label B'. Set your compass to the length of $\overline{BC}$. Place the compass point at B' and mark a second point C'. $\overline{B'C'}$ has the same length as $\overline{BC}$. Now place the compass point at B'. Use a compass setting greater than half the length of $\overline{B'C'}$ and draw an arc above and below $\overline{B'C'}$. Using the same compass setting with the point at C', draw an arc above and below $\overline{B'C'}$. Use a straightedge to draw a segment through the points of arc intersection. This segment intersects $\overline{B'C'}$ at point D. Segment $\overline{B'D}$ has the same length as segment $\overline{DC'}$.

b. Bisect angle $\angle C$.

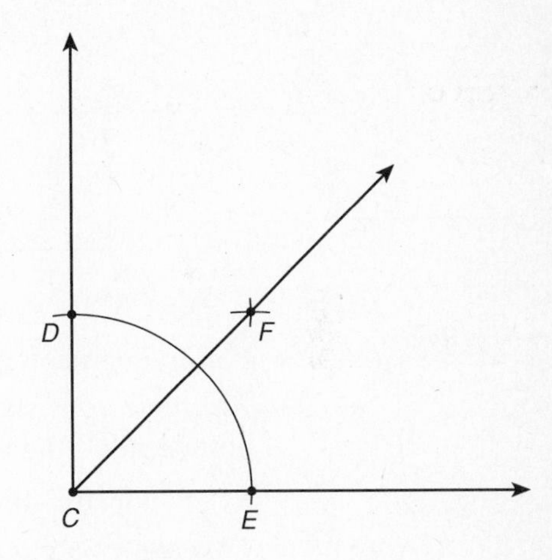

Place the compass point at vertex C and draw an arc that intersects both rays of $\angle C$. Label the points D and E. Place the compass point at D and draw an arc in the interior of the angle $\angle C$. Using the same compass setting, place the point at E and draw a second arc in the interior of $\angle C$. Label the point of intersection as F. Use a straightedge to draw the ray from C through point F.

Guidelines:

- To *bisect a segment* means to find the midpoint of the segment, the point that divides the segment into congruent segments.
- To *bisect an angle* means to find the ray that divides it into two angles, each of which is half the measure of the original angle.

EXERCISES

1. Draw an angle that measures about 30°. Copy it, then bisect it.

2. Draw a segment that is about 1 inch long. Copy it, then bisect it.

Reteach
Chapter 2

Name ____________________

What you should learn:

2.1 How to identify patterns and how to read definitions

Correlation to Pupil's Textbook:

Mid-Chapter Self-Test (p. 77)
Exercises 1, 2, 4, 7, 9, 10, 12, 17, 20

Chapter Test (p. 101)
Exercises 2, 4, 7, 9, 10

Examples *Identifying Patterns and Reading Definitions*

a. Find a pattern and a formula for the nth number.

The pattern is 1, 4, 9, 16, 25, 36. These are called square numbers. The formula for the nth number is n^2.

n	1	2	3	4	5	6
nth Number	1	4	9	?	?	?

b. Draw each of the following: $\overleftrightarrow{CD}$, $\overrightarrow{CD}$, $\overline{CD}$, and $\overrightarrow{DC}$.

$\overleftrightarrow{CD}$

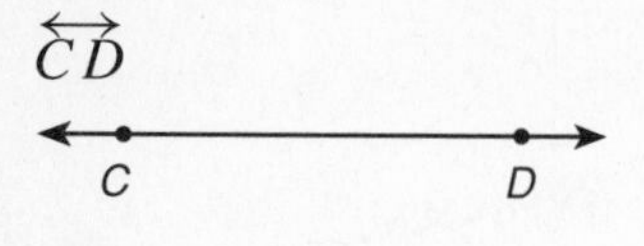

$\overrightarrow{CD}$

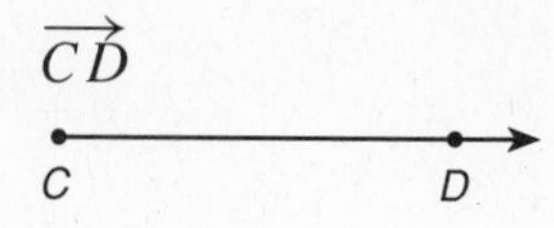

$\overline{CD}$

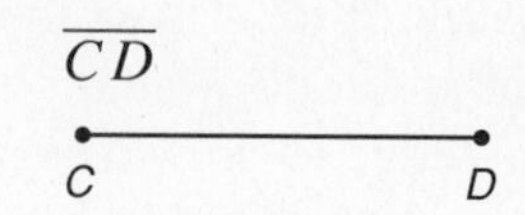

$\overrightarrow{DC}$

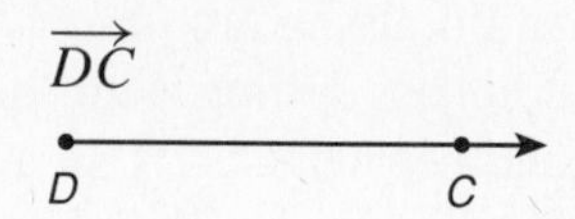

c. Draw the following angles: acute, right, obtuse, straight.

Acute
($0° < m\angle x < 90°$)

Right
($m\angle x = 90°$)

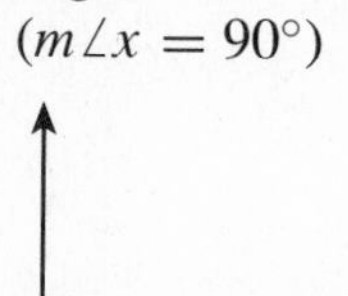

Obtuse
($90° < m\angle x < 180°$)

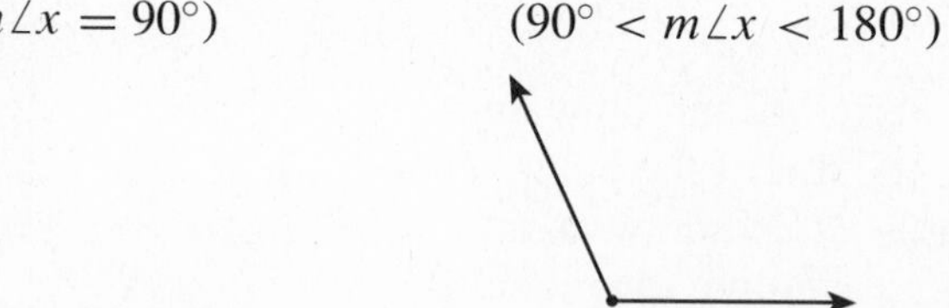

Straight
($m\angle x = 180°$)

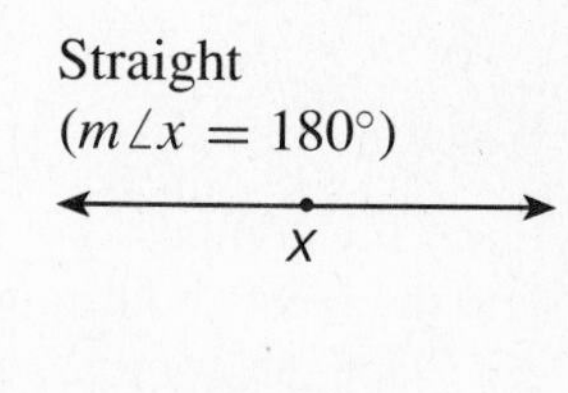

Guidelines:

- Points, segments, or rays that are on the same line are *collinear*.
- An *angle* consists of two different rays that have the same initial point (called the *vertex*). The rays are the sides of the angle.
- Every nonstraight angle has an *interior* and an *exterior*.
- Two angles are *adjacent* if they share a common vertex and side, but have no common interior points.

EXERCISES

In Exercises 1–5, use the figure at the right to identify each of the following.

1. Opposite rays
2. Acute angle
3. Obtuse angle
4. Adjacent angles
5. Straight angle

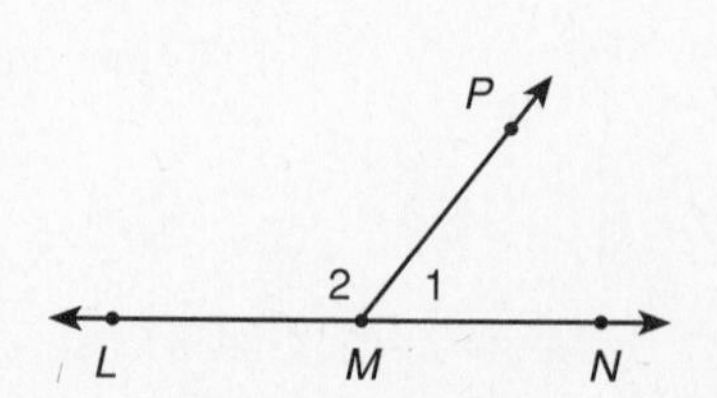

6. Find a pattern and a formula for the nth number.

n	1	2	3	4	5	6
nth Number	3	6	9	?	?	?

Reteach
Chapter 2

Name ______________________________

What you should learn:

2.2	How to use segment postulates and angle postulates

Correlation to Pupil's Textbook:

Mid-Chapter Self-Test (p. 77) Exercises 5, 11, 18, 19

Chapter Test (p. 101) Exercise 14

Examples *Using Segment and Angle Postulates*

a. Find LM, on the line to the right, using the Ruler Postulate.

Answer: $LM = |5 - 8| = 3$ or $LM = |8 - 5| = 3$

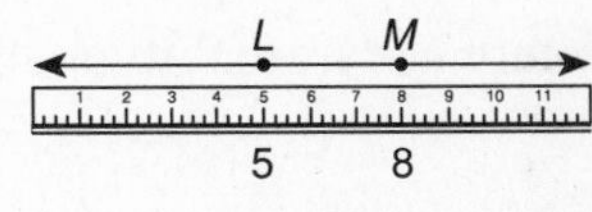

b. If Q is between C and D, find QD.

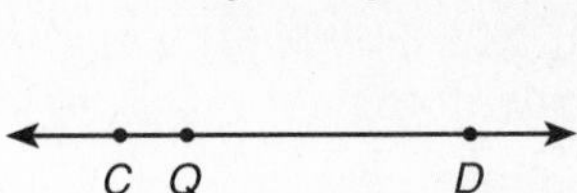

By the Segment Addition Postulate, $CQ + QD = CD$. To find QD, subtract CQ from both sides of the equation. Then, $QD = CD - CQ$.

c. Using the figure below, find $m\angle OMN$.

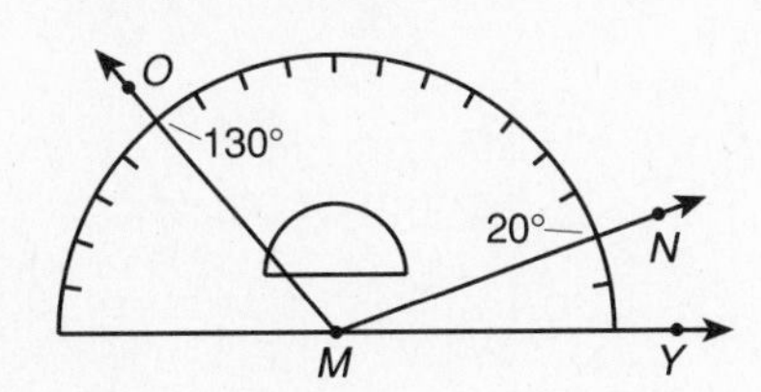

Answer: By the Protractor Postulate,

$$m\angle OMN = m\angle OMY - m\angle NMY$$
$$= 130 - 20$$
$$= 110°.$$

d. Using the figure below, find $m\angle QPS$.

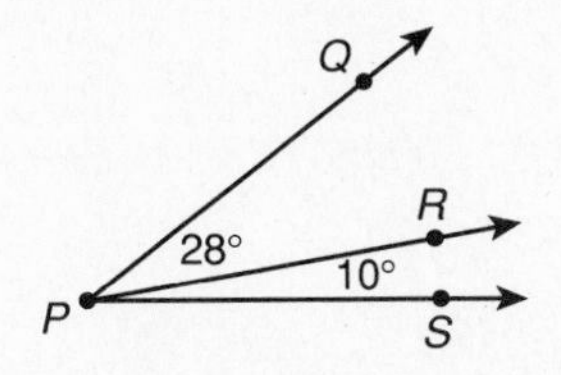

Answer: By the Angle Addition Postulate,

$$m\angle QPS = m\angle QPR + m\angle RPS$$
$$= 28 + 10$$
$$= 38°.$$

Guidelines:

- *Postulates* or *axioms* are statements that you accept without proof.
- *Theorems* are statements that you can prove.
- If B is *between* A and C, then $AB + BC = AC$.
- If C is in the *interior* of $\angle AOD$, then $m\angle AOC + m\angle COD = m\angle AOD$.

EXERCISES

In Exercises 1–3, use the figure at the right and the following information: $CD = EF$, $CF = 10$, $DF = 8$.

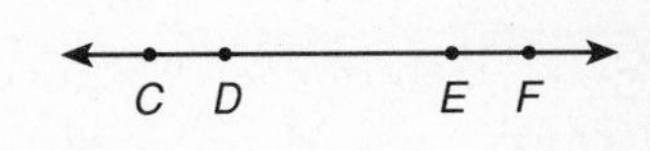

1. Find CD.

2. Find EF.

3. Find CE.

4. Draw adjacent angles $\angle RST$ and $\angle TSV$. If $m\angle RST = 85°$ and $m\angle RSV = 125°$, find $m\angle TSV$.

Reteach
Chapter 2

Name ______________________________

What you should learn:

2.3	How to relate segments and angles and solve real-life problems

Correlation to Pupil's Textbook:

Mid-Chapter Self-Test (p. 77)
Exercises 3, 6, 8, 13–17, 20

Chapter Test (p. 101)
Exercises 1, 2, 5, 8, 11, 13

Examples *Building Geometric Vocabulary and Solving Real-life Problems*

a. Use the figure at the right to find the indicated points, rays, angles, and segments.

	Answers:
1. Congruent segments	1. $\overline{CD} \cong \overline{CF}$, $\overline{DE} \cong \overline{FE}$
2. Congruent angles	2. $\angle DEC \cong \angle FEC$, $\angle DCE \cong \angle FCE$
3. Midpoint of $\overline{DF}$	3. E
4. Segment bisector of $\overline{DF}$	4. $\overline{CE}$, $\overrightarrow{CE}$, or $\overleftrightarrow{CE}$
5. Angle bisector of $\angle DCF$	5. $\overrightarrow{CE}$
6. Perpendicular lines	6. $\overleftrightarrow{CE} \perp \overleftrightarrow{DF}$

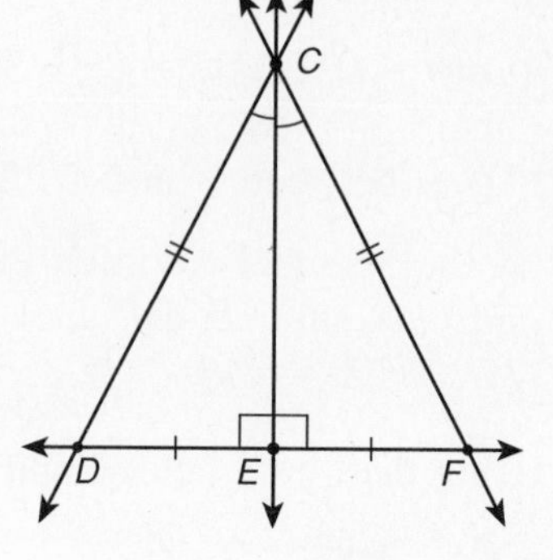

b. Use the Distance Formula to find the distance between the airport and the city. Each unit on the graph equals one mile.

Let $A = (x_1, y_1) = (-3, -1)$ and $C = (x_2, y_2) = (4, 5)$.

$$AC = \sqrt{(x_2 - x_1)^2 + (y_2 - y_1)^2}$$
$$= \sqrt{(4 - (-3))^2 + (5 - (-1))^2}$$
$$= \sqrt{7^2 + 6^2}$$
$$= \sqrt{49 + 36} = \sqrt{85} \approx 9 \text{ miles}$$

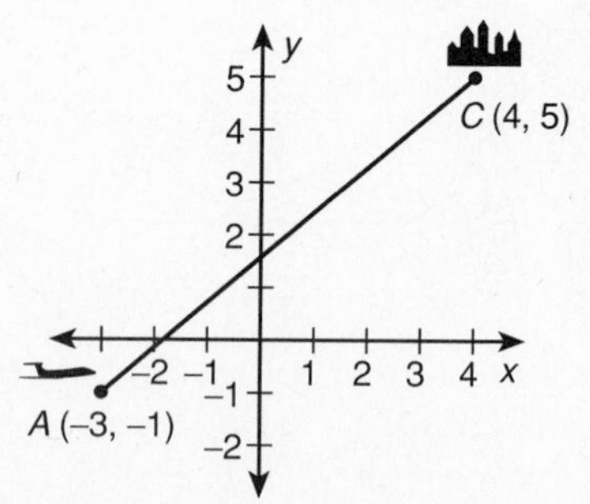

Guidelines:

- *Congruent segments* have the same length and *congruent angles* have the same measure.
- The *midpoint* of a segment is the point that divides the segment into two congruent segments.
- Two lines are *perpendicular* ($\perp$) if they intersect to form four right angles.

EXERCISES

Let $M = (-2, 3)$ and $N = (1, -1)$.

1. Find MN.

2. Find the midpoint of $\overline{MN}$.

In Exercises 3–5, choose the word (always, sometimes, or never) that best completes the statement.

3. If two lines are perpendicular, they [?] intersect to form right angles.

4. If two lines intersect to form right angles, they are [?] perpendicular.

5. If a line is perpendicular to a segment, it [?] is a segment bisector.

Reteach
Chapter 2

Name ________________________________

What you should learn:

2.4	How to use conditional statements and postulates which are conditional statements

Correlation to Pupil's Textbook:

Chapter Test (p. 101)
Exercises 17–19

Examples *Using Conditional Statements and Using Postulates*

a. State the hypothesis, conclusion, and converse of this conditional statement: If it is raining, then the tennis match will be postponed.

Hypothesis: It is raining.
Conclusion: The tennis match will be postponed.
Converse: If the tennis match is postponed, then it is raining.

b. Write the following statement in "if and only if" form. A straight angle is an angle that measures 180°.

An angle is a straight angle if and only if its measure is 180°.

c. State the converse of the given conditional statement. Then show, with a counterexample, that the converse is false.
Statement: If $\angle A$ and $\angle B$ are right angles, then $\angle A \cong \angle B$.

Converse: If $\angle A \cong \angle B$, then $\angle A$ and $\angle B$ are right angles.
Counterexample: Let $m\angle A = m\angle B = 60°$. Then $\angle A \cong \angle B$, but $\angle A$ and $\angle B$ are not right angles.

Guidelines:

- A *conditional statement* has a hypothesis, p, and a conclusion, q, stated "If p, then q."
- The *converse* of a conditional statement is stated "If q, then p."
- A *biconditional statement,* "p if and only if q," is equivalent to "if p, then q" and "if q, then p."

EXERCISES

In Exercises 1–4, state the hypothesis, conclusion, and converse of each conditional statement.

1. If $m\angle C = 100°$, then $\angle C$ is obtuse.
2. If points K and L are different points in a plane, then there is a third point in the plane not on $\overleftrightarrow{KL}$.
3. If two distinct planes intersect, then their intersection is a line.
4. If points P, Q, and R are noncollinear, then they lie in one and only one plane.
5. Write the following statement in "if and only if" form.
   Two lines that intersect to form right angles are defined to be perpendicular.

Reteach
Chapter 2

Name ______________________

What you should learn:

2.5	How to use properties from algebra and properties of congruence

Correlation to Pupil's Textbook:

Chapter Test (p. 101)
Exercises 15, 16

Examples *Using Properties from Algebra and Properties of Congruence*

a. Match the statement on the left with the Property of Equality on the right.

1. If $m\angle 1 = m\angle 2$, then $m\angle 1 + m\angle 3 = m\angle 2 + m\angle 3$.
2. If $x = TV$, then $TV = x$.
3. If $m\angle 4 = 40°$, then $\frac{1}{2}m\angle 4 = 20°$.
4. If $AB = CD$ and $CD = EF$, then $AB = EF$.

a. Addition Property of Equality
b. Transitive Property of Equality
c. Symmetric Property of Equality
d. Multiplication Property of Equality

Answers: 1. a 2. c 3. d 4. b

b. Match the statement on the left with the Property of Congruence on the right.

1. If $\angle E \cong \angle F$, then $\angle F \cong \angle E$.
2. $\overline{PQ} \cong \overline{PQ}$
3. If $\angle G \cong \angle H$ and $\angle H \cong \angle J$, then $\angle G \cong \angle J$.

a. Reflexive Property of Congruence
b. Transitive Property of Congruence
c. Symmetric Property of Congruence

Answers: 1. c 2. a 3. b

Guidelines:

- *Properties of equality* are used with segment lengths and angle measures because they are real numbers.
- *Properties of congruence* are used with geometric objects that have the same size and the same shape.

EXERCISES

In Exercises 1–5, match the correct reason for each step in the proof.

Given: $m\angle 1 = m\angle 2$
Prove: $m\angle RSU = m\angle TSV$

1. $m\angle 1 = m\angle 2$
2. $m\angle 1 + m\angle TSU = m\angle 2 + m\angle TSU$
3. $m\angle RSU = m\angle 1 + m\angle TSU$
4. $m\angle TSV = m\angle 2 + m\angle TSU$
5. $m\angle RSU = m\angle TSV$

a. Angle Addition Postulate
b. Given
c. Substitution Property
d. Addition Prop. of Equality
e. Angle Addition Postulate

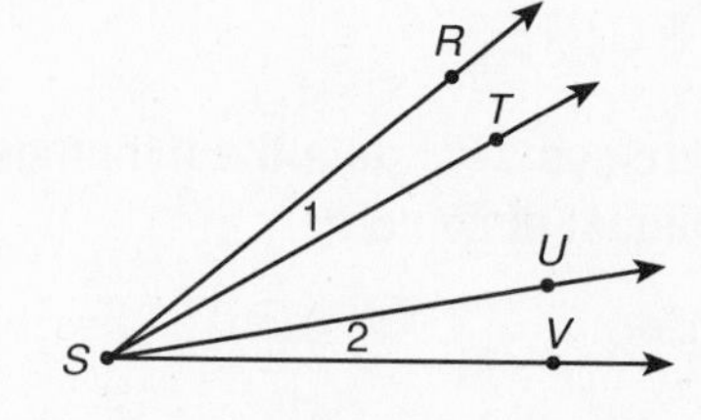

In Exercises 6–9, match the statement with the property on the right.

6. $\overline{AB} \cong \overline{AB}$
7. If $\angle B \cong \angle C$, then $\angle C \cong \angle B$.
8. If $m\angle X = m\angle Y$ and $m\angle Y = m\angle Z$, then $m\angle X = m\angle Z$.
9. $XY = XY$

a. Symmetric Property of Congruence
b. Transitive Property of Equality
c. Reflexive Property of Equality
d. Reflexive Property of Congruence.

Reteach
Chapter 2

Name ______________________________

What you should learn:

2.6 How to identify special pairs of angles

Correlation to Pupil's Textbook:

Chapter Test (p. 101)
Exercises 3, 6, 12

Examples *Identifying Special Pairs of Angles*

a. Two angles are vertical angles if the sides form two pairs of opposite rays. Two angles are complementary if the sum of their measures is 90°. (Each angle is the complement of the other.)
In the figure at the right, name two labeled angles that are vertical angles.

Answer: $\angle 1$ and $\angle 3$ are vertical angles.

In the figure at the right, name two labeled angles that are complementary angles.

Answer: $\angle 4$ and $\angle 5$ are complementary angles.

b. Use the figure at the right to find the indicated pairs of angles.

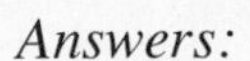

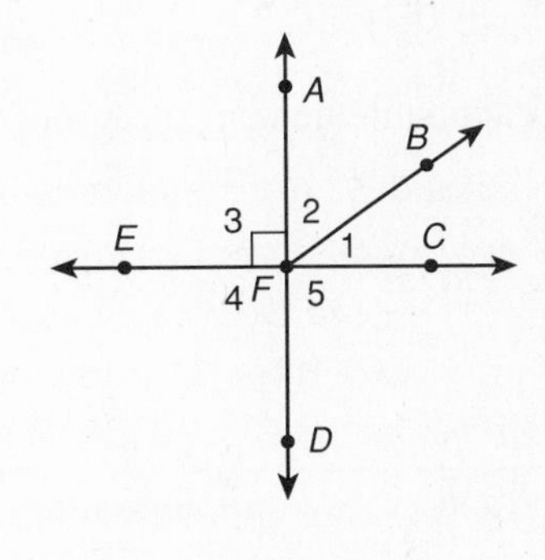

	Answers:
1. Vertical angles	$\angle 3$ and $\angle 5$, $\angle 4$ and $\angle AFC$
2. Linear pair of angles (also supplementary)	$\angle 3$ and $\angle 4$
	$\angle 4$ and $\angle 5$
	$\angle 5$ and $\angle CFA$
	$\angle CFA$ and $\angle 3$
3. Complementary angles	$\angle 1$ and $\angle 2$

Guidelines:

- Two adjacent angles are a *linear pair* if their noncommon sides are opposite rays.
- If two angles form a linear pair, then they are supplementary.

EXERCISES

In Exercises 1–4, sketch a pair of angles that fits the description.

1. Adjacent congruent complementary angles
2. Congruent vertical angles
3. A linear pair of angles
4. Supplementary angles
5. Sketch, if possible, non-adjacent, congruent supplementary angles.
6. Sketch, if possible, non-congruent vertical angles.

Reteach
Chapter 2

Name ____________________

What you should learn:

2.6	How to use deductive reasoning to verify angle relationships

Correlation to Pupil's Textbook:

Chapter Test (p. 101)
Exercise 20

Examples *Using Deductive Reasoning to Verify Angle Relationships*

a. Give a reason for each step in the proof.

Given: $\overrightarrow{OL} \perp \overrightarrow{ON}$
Prove: $m\angle 1 = 90° - m\angle 2$
Proof:

1. $\overrightarrow{OL} \perp \overrightarrow{ON}$ — 1. Given
2. $\angle LON$ is a right angle. — 2. If two lines are $\perp$, then they intersect to form a right angle.
3. $m\angle LON = 90°$ — 3. Definition of a right angle
4. $m\angle 1 + m\angle 2 = m\angle LON$ — 4. Angle Addition Postulate
5. $m\angle 1 + m\angle 2 = 90°$ — 5. Substitution Property of Equality
6. $m\angle 1 = 90° - m\angle 2$ — 6. Subtraction Property of Equality

b. Match each statement in the proof below with the correct reason.

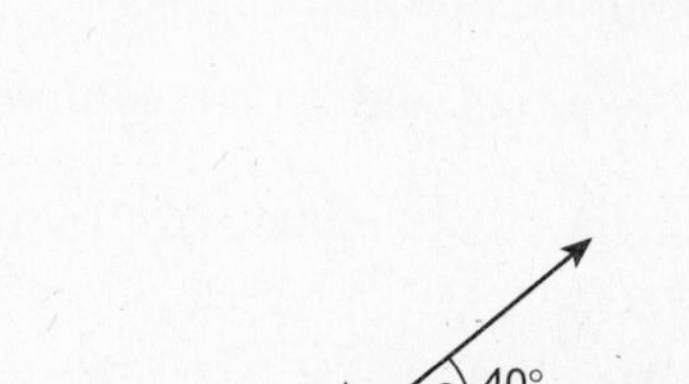

Given: $\angle 1$ and $\angle 2$ are supplementary angles.
$m\angle 2 = 40°$
Prove: $m\angle 1 = 140°$
Proof:

Statements	*Reasons*
1. $\angle 1$ and $\angle 2$ are supplementary angles.	?
2. $m\angle 1 + m\angle 2 = 180°$	?
3. $m\angle 2 = 40°$	?
4. $m\angle 1 + 40° = 180°$	?
5. $m\angle 1 = 140°$	?

a. Def. of supplementary angles
b. Given
c. Given
d. Subtraction Property of Equality
e. Substitution Property of Equality

Answers: 1. b or c 2. a 3. b or c 4. e 5. d

Guidelines:

- If two angles are supplementary to the same angle or congruent angles, then they are congruent.
- If two angles are complementary to the same angle or congruent angles, then they are congruent.
- If two angles are vertical angles, then they are congruent.

EXERCISES

1. Prove if two angles are complementary to the same angle, then they are congruent.

2. Prove if two angles are complementary to congruent angles, then they are congruent.

Reteach
Chapter 3

Name ______________________

What you should learn:

3.1	How to identify parallel and perpendicular lines and planes

Correlation to Pupil's Textbook:

Mid-Chapter Self-Test (p. 133)
Exercises 1–4

Chapter Test (p. 157)
Exercises 1–4

Examples *Identifying Parallel and Perpendicular Lines and Planes*

a. In the figure at the right, $\ell_1 \parallel \ell_2$ and $\ell_1 \parallel \ell_3$. How are lines ℓ_2 and ℓ_3 related?

By the Transitive Property of Parallel Lines, $\ell_2 \parallel \ell_3$. Although not drawn, there is a plane that contains both ℓ_2 and ℓ_3.

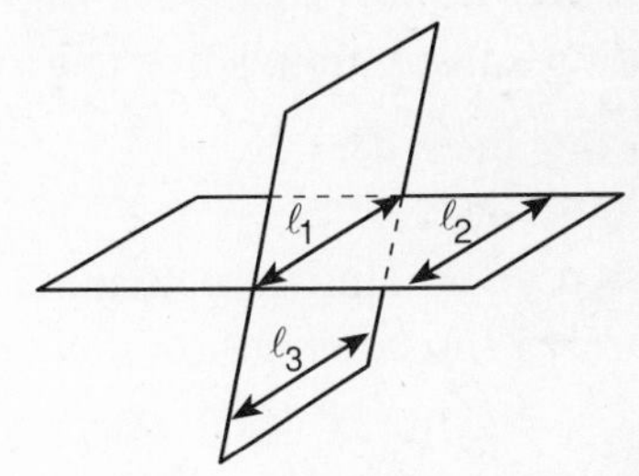

b. In the figure at the right, $\ell_1 \perp \ell_2$ and $\ell_2 \perp \ell_3$. How are lines ℓ_1 and ℓ_3 related?

Since ℓ_1 and ℓ_3 are coplanar, by the Property of Perpendicular Lines Theorem, ℓ_1 is parallel to ℓ_3.

How are lines ℓ_2 and ℓ_4 related?

Since ℓ_2 and ℓ_4 are not in the same plane, they are skew lines.

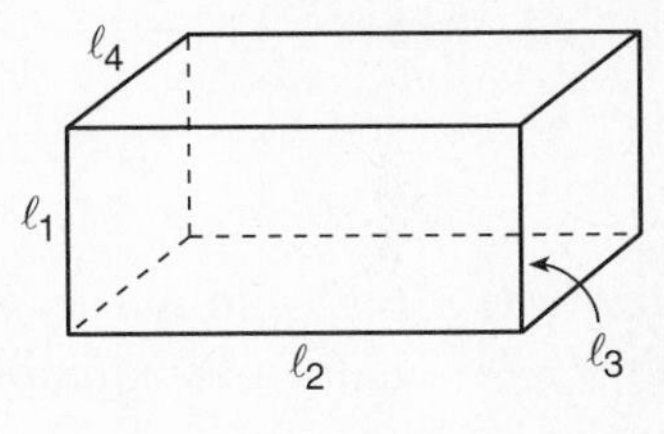

Guidelines:

- *Parallel lines* are coplanar lines that do not intersect.
- *Intersecting lines* are coplanar and have exactly one point in common. They are *perpendicular* if they meet at right angles; otherwise, they are *oblique*.
- *Skew lines* are non-coplanar lines. They do not intersect.
- From algebra, two nonvertical lines are *parallel* if and only if they have the same slope.
- From algebra, two nonvertical lines are *perpendicular* if and only if the product of their slopes is -1.

EXERCISES

In Exercises 1–5, use the figure at the right to complete each statement.

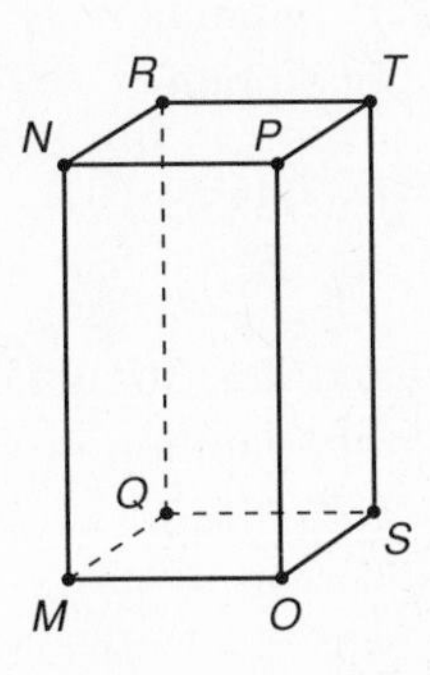

1. If $\overleftrightarrow{MO} \parallel \overleftrightarrow{QS}$ and $\overleftrightarrow{QS} \parallel \overleftrightarrow{RT}$, then $\boxed{?} \parallel \boxed{?}$.
2. If $\overleftrightarrow{MO} \perp \overleftrightarrow{OP}$ and $\overleftrightarrow{OP} \perp \overleftrightarrow{NP}$, and $\overleftrightarrow{MO}$, $\overleftrightarrow{OP}$, and $\overleftrightarrow{NP}$ are coplanar, then $\boxed{?} \parallel \boxed{?}$.
3. Plane MOS and plane NPT are $\boxed{?}$.
4. Plane MOS and plane SOP are $\boxed{?}$.
5. Lines $\overleftrightarrow{MQ}$ and $\overleftrightarrow{NP}$ are $\boxed{?}$.

Reteach
Chapter 3

Name ______________________

What you should learn:

3.2 How to solve a system of linear equations

Correlation to Pupil's Textbook:

Mid-Chapter Self-Test (p. 133) Exercise 6

Chapter Test (p. 157) Exercise 5

Examples *Solving Linear Systems*

a. Using substitution, solve the following linear system.

$$\begin{cases} 4x - 5y = 13 \\ 3x - y = 7 \end{cases}$$

From your solution, determine if the lines are intersecting (one solution), parallel (no solution), or coincident (many solutions).

$3x - 7 = y$	*Solve the second equation for y.*
$4x - 5(3x - 7) = 13$	*Substitute into the first equation.*
$4x - 15x + 35 = 13$	*Simplify.*
$-11x = -22$	*Simplify.*
$x = 2$	*Divide both sides by −11.*

Substitute $x = 2$ into the second equation to obtain $y = -1$. The ordered pair $(2, -1)$ is the one solution, thus, the lines are intersecting.

b. Using substitution, solve the following linear system.

$$\begin{cases} x + 3y = 6 \\ 2x + 6y = 10 \end{cases}$$

From your solution, determine if the lines are intersecting (one solution), parallel (no solution), or coincident (many solutions).

$x = 6 - 3y$	*Solve the first equation for x.*
$2(6 - 3y) + 6y = 10$	*Substitute into the second equation.*
$12 - 6y + 6y = 10$	*Simplify.*
$12 = 10$	*Simplify.*

There is no solution for this system, therefore, the lines are parallel.

Guidelines:

- A system of two linear equations can have exactly one solution, no solution, or infinitely many solutions.
- If two distinct lines intersect, then their intersection is exactly one point.

EXERCISES

1. Solve the system and determine if the following lines are intersecting, parallel, or coincident.

$$\begin{cases} x - 2y = -6 \\ \frac{1}{2}x - y = -3 \end{cases}$$

2. Determine if the following lines are intersecting, parallel, or coincident.

$$\begin{cases} 3x + 9y = 8 \\ 2x + 6y = 7 \end{cases}$$

Reteach

Chapter 3

Name ______________________________

What you should learn:

3.2 How to find an equation of a line

Correlation to Pupil's Textbook:

Mid-Chapter Self-Test (p. 133) Exercises 5, 6

Chapter Test (p. 157) Exercises 5, 6

Examples *Finding an Equation of a Line*

a. The line ℓ_1 is given by $6x - 2y = 3$. Find an equation for ℓ_2 passing through the point (2, 1) and parallel to ℓ_1.

Rewriting the equation for ℓ_1 in slope-intercept form, you can determine ℓ_1 has a slope of 3. Because $\ell_1 \parallel \ell_2$, ℓ_2 has the same slope.

$y = mx + b$	*Slope-intercept form*
$1 = 3(2) + b$	*Substitute 3 for m, 2 for x, and 1 for y.*
$1 = 6 + b$	*Simplify.*
$-5 = b$	*Solve for the y-intercept, b.*
$y = 3x - 5$	*Equation for ℓ_2*

b. The line ℓ_1 is given by $2x + 3y = 4$. Find an equation for ℓ_2 passing through the point (2, −2) and perpendicular to ℓ_1.

Rewriting the equation for ℓ_1 in slope-intercept form, you can determine ℓ_1 has a slope of $-\frac{2}{3}$. Because ℓ_2 is perpendicular to ℓ_1, ℓ_2 has a slope of $\frac{3}{2}$.

$y = mx + b$	*Slope-intercept form*
$-2 = \left(\frac{3}{2}\right)(2) + b$	*Substitute $\frac{3}{2}$ for m, 2 for x, and −2 for y.*
$-2 = 3 + b$	*Simplify.*
$-5 = b$	*Solve for the y-intercept, b.*
$y = \frac{3}{2}x - 5$	*Equation for ℓ_2*

Guidelines:

- If there is a line and a point not on the line, then there is exactly one line through the point parallel to the given line (Parallel Postulate).
- If there is a line and a point not on the line, then there is exactly one line through the point perpendicular to the given line (Perpendicular Postulate).

EXERCISES

1. The line ℓ_1 is given by $5x - 2y = 7$. Find an equation for ℓ_2 passing through the point (−2, 3) and perpendicular to ℓ_1.

2. Write an equation of the line that is parallel to ℓ and passes through P, as shown at the right.

   ℓ: $y = -\frac{1}{2}x + 2$

   P: (4, 6)

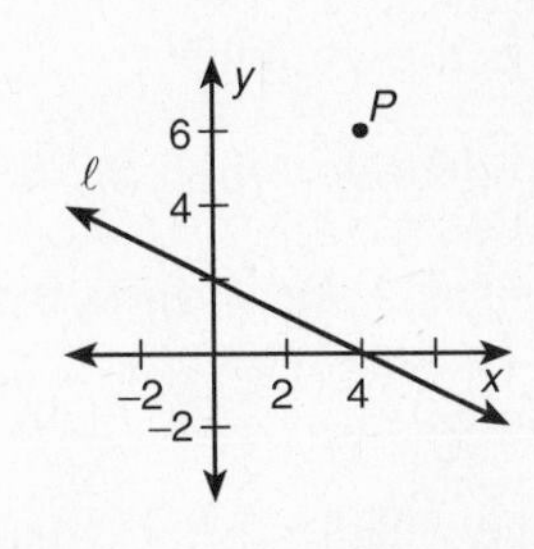

Reteach

Chapter 3

Name ______________________________

What you should learn:

3.3 How to form the negation and contrapositive of a conditional statement

Correlation to Pupil's Textbook:

Mid-Chapter Self-Test (p. 133) Exercises 7–8

Chapter Test (p. 157) Exercise 11

Examples *Using Contrapositives*

a. Write the contrapositive and the converse of the conditional statement.

Original statement: $p \Rightarrow q$
If school is closed on Monday, then it is a holiday.

Contrapositive: $\sim q \Rightarrow \sim p$
If it is not a holiday, then the school is not closed on Monday.

Converse: $q \Rightarrow p$
If it is a holiday, then school is closed on Monday.

b. When is the contrapositive true?

The contrapositive is true if and only if the conditional statement is true.

c. Write the following statement in "if-then" form.

Vertical angles are congruent.

If two angles are vertical angles, then they are congruent.

d. Write the contrapositive and the converse of the "if-then" statement in Example c.

Contrapositive: $\sim q \Rightarrow \sim p$
If two angles are not congruent, then they are not vertical angles.

Converse: $q \Rightarrow p$
If two angles are congruent, then they are vertical angles.

Guidelines:

- The *negation* of a hypothesis or of a conclusion is formed by denying the original hypothesis or conclusion.
- The *contrapositive* of the conditional statement $p \Rightarrow q$ is $\sim q \Rightarrow \sim p$.

EXERCISES

In Exercises 1 and 2, write the negation of the statement.

1. $\angle 1$ is a right angle.

2. The measure of $\angle 1$ is 90°.

In Exercises 3 and 4, write the contrapositive of the conditional statement.

3. If Jared earns \$150, then he will buy a compact disc player.

4. If the weather is too windy, then the sailing race will be canceled.

Reteach
Chapter 3

Name ______________________________

What you should learn:

3.3 How to use a syllogism to reason deductively

Correlation to Pupil's Textbook:

Mid-Chapter Self-Test (p. 133)
Exercise 9

Chapter Test (p. 157)
Exercises 12–13

Examples *Using Syllogisms to Reason Deductively*

a. Assume the statements are true. Write the conclusion of the syllogism.

If the soccer team wins the last game, then the school will win the championship. $p \Rightarrow q$

If the school wins the championship, then a new trophy will be placed in the showcase. $q \Rightarrow r$

Conclusion: If the soccer team wins the last game, then a new trophy will be placed in the showcase. $p \Rightarrow r$

b. Assume the statements are true. Write the conclusion, using the Law of Detachment.

If the senior class earns at least $1200 from selling magazine subscriptions, then the class will visit Niagara Falls. $p \Rightarrow q$

The senior class earns $1275. p is true.

Conclusion: The class will visit Niagara Falls. q is true.

Guidelines:

- The Law of Syllogism: If $p \Rightarrow q$ and $q \Rightarrow r$, then $p \Rightarrow r$.
- The Law of Detachment: If $p \Rightarrow q$ and p is true, then q is true.

EXERCISES

In Exercises 1 and 2, state the law of logic which is illustrated by the following statements.

1. If Ann is the oldest daughter, then Ann has the largest bedroom. Ann is the oldest daughter. Ann does have the largest bedroom.

2. If $m\angle A = 40°$, then $m\angle B = 140°$.
If $m\angle B = 140°$, then $m\angle C = 140°$.
If $m\angle A = 40°$, then $m\angle C = 140°$.

In Exercises 3 and 4, assume the statements are true. Write the conclusion and state the law of logic which is illustrated.

3. If it snows today, then we will go skiing. It snows today.

4. If it snows today, then our family will go skiing. If our family goes skiing, then we will need motel reservations. It snows today.

Reteach
Chapter 3

Name ____________________

What you should learn:

3.4 How to read and write different styles of proofs

Correlation to Pupil's Textbook:

Mid-Chapter Self-Test (p. 133)
Exercises 10–12

Chapter Test (p. 157)
Exercises 19, 20

Examples *Using Different Styles of Proofs*

a. State the reason for each step in the two-column proof.

Given: $\angle 1$ and $\angle 2$ are supplementary.
$\angle 2 \cong \angle 3$

Prove: $m\angle 1 + m\angle 3 = 180°$

Proof:

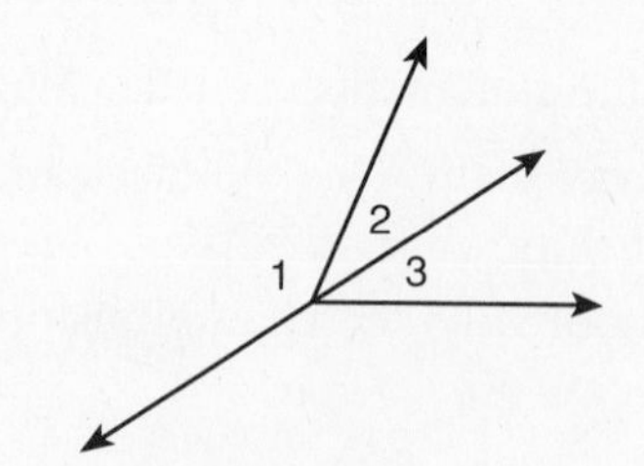

Statements	*Reasons*
1. $\angle 1$ and $\angle 2$ are supplementary.	1. ? Given
2. $m\angle 1 + m\angle 2 = 180°$	2. ? Def. of supplementary angles
3. $\angle 2 \cong \angle 3$	3. ? Given
4. $m\angle 2 = m\angle 3$	4. ? Def. of congruent angles
5. $m\angle 1 + m\angle 3 = 180°$	5. ? Substitution Property of Equality

b. Write a two-column proof for Theorem 3.3–If two lines are perpendicular, then they intersect to form four right angles. Use the diagram at the right.

Given: $\ell_1 \perp \ell_2$

Prove: $m\angle 1 = m\angle 2 = m\angle 3 = m\angle 4 = 90°$

Proof:

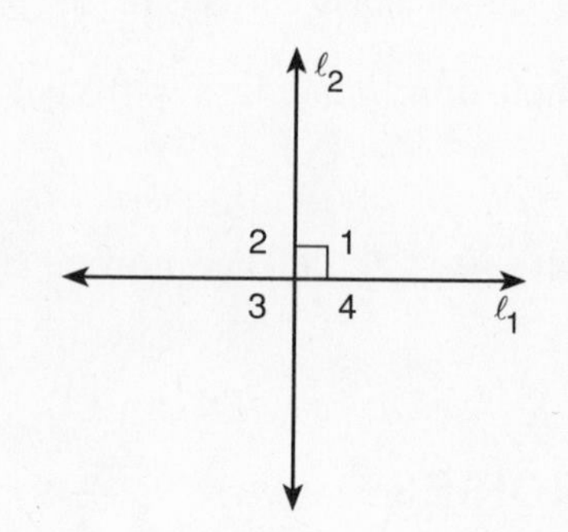

Statements	*Reasons*
1. $\ell_1 \perp \ell_2$	1. Given
2. $\angle 1$ is a right angle.	2. Def. of perpendicular lines
3. $m\angle 1 = 90°$	3. Def. of right angle
4. $\angle 1 \cong \angle 3$	4. Vertical Angles Theorem
5. $m\angle 1 = m\angle 3 = 90°$	5. Def. of congruent angles
6. $\angle 1$ and $\angle 2$ are supplementary. $\angle 3$ and $\angle 4$ are supplementary.	6. Linear Pair Postulate
7. $m\angle 1 + m\angle 2 = 180°$ $m\angle 3 + m\angle 4 = 180°$	7. Def. of supplementary angles
8. $m\angle 2 = 90°$ $m\angle 4 = 90°$	8. Subtraction Property of Equality

Name ____________________

c. Write a flow proof, using the diagram at the right.

Given: $m\angle 1 = m\angle 3$

Prove: $m\angle QPS = m\angle TPR$

Proof:

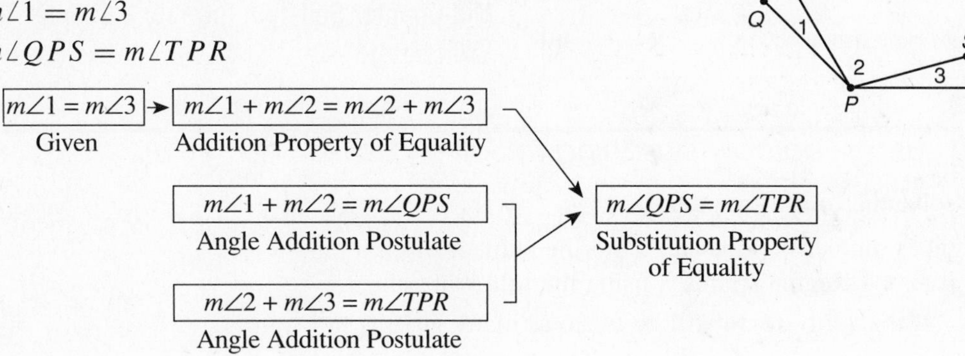

d. Write a paragraph proof for Theorem 3.4–All right angles are congruent.

Given: $\angle 1$ and $\angle 2$ are right angles

Prove: $\angle 1 \cong \angle 2$

- Because $\angle 1$ and $\angle 2$ are right angles, you know that $m\angle 1 = 90°$ and $m\angle 2 = 90°$.
- By the substitution property of equality, $m\angle 1 = m\angle 2$.
- Using the definition of congruent angles, you can conclude that $\angle 1 \cong \angle 2$.

Guidelines:

- A *paragraph proof* is written as a narrative.
- A *flow proof* uses arrows to show the flow of the logical argument and has reasons written below each statement.
- A *two-column proof* is a sequence of numbered statements and reasons.

EXERCISES

1. Rework Example c by writing a paragraph proof.
2. Rework Example d by writing a two-column proof.
3. Match the correct reason for each step in the proof.

Given: $\angle R$ and $\angle S$ are complementary.
$\angle S$ and $\angle T$ are complementary.

Prove: $\angle R \cong \angle T$

Proof:

Statements	*Reasons*
1. $\angle R$ and $\angle S$ are complementary.	?
2. $m\angle R + m\angle S = 90°$	?
3. $\angle S$ and $\angle T$ are complementary.	?
4. $m\angle S + m\angle T = 90°$	?
5. $m\angle R + m\angle S = m\angle S + m\angle T$	?
6. $m\angle R = m\angle T$	?
7. $\angle R \cong \angle T$	?

a. Def. of complementary angles
b. Substitution Prop. of Equality
c. Given
d. Given
e. Def. of congruent angles
f. Def. of complementary angles
g. Subtraction Prop. of Equality

Reteach

Chapter 3

Name ______________________

What you should learn:

3.4 How to use logical reasoning

Correlation to Pupil's Textbook:

Mid-Chapter Self-Test (p. 133) Exercises 10–12

Chapter Test (p. 157) Exercises 19, 20

Examples *Using Logical Reasoning to Solve Puzzles*

a. Use logic to solve the following puzzle.

Tom and Regina's mother is pregnant. Find the number of children presently in Tom and Regina's family, using the following clues.

1. Tom says, "If it's a girl, there will be twice as many girls as there are boys."
2. Regina says, "If it's a boy, there will be as many boys as there are girls."

From clue 1, the girls in the family plus a baby girl equals twice the number of boys. In symbols, $g + 1 = 2b$.

From clue 2, the boys in the family plus a baby boy equals the number of girls. In symbols, $b + 1 = g$.

Try solving the system of equations.

$$\begin{cases} g - 2b = -1 \\ -g + b = -1 \end{cases}$$

Solving the system of equations gives the solution of 2 boys and 3 girls presently in the family.

b. Use logic to solve the following puzzle.

Six runners are participating in a 400-meter race. As the winner crosses the finish line, the following conditions exist.

1. Don is 25 meters behind Cara.
2. Cara is 15 meters ahead of Stacie.
3. Mark is running alongside Jonetra.
4. Stacie is 30 meters behind Ramon, who is 5 meters ahead of Jonetra.

List the order of the runners when the winner crosses the finish line.

You can place the runners on a number line to determine the order at the finish line as shown at the right.

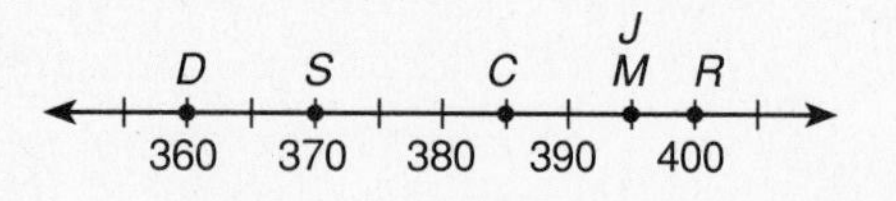

Ramon is the winner, followed by Mark and Jonetra, followed by Cara, Stacie, and Don.

EXERCISE

Five friends compared their vehicles. Two of them drove trucks, while three of them drove automobiles. Additional facts are as follows:

1. Luke and Dedra had different vehicles.

2. Carlo and Amber had the same vehicles.

3. Amber and Luke had different vehicles.

4. The name of the fifth friend was Juan.

Name the friends who drove trucks and those who drove automobiles.

Reteach
Chapter 3

Name ______________________________

What you should learn:

3.5	How to name angles formed by a transversal and how to use properties of parallel lines

Correlation to Pupil's Textbook:

Chapter Test (p. 157)
Exercises 7–10

Examples *Naming Angles and Using Properties of Parallel Lines*

a. Using the figure at the right, name the angles formed by the transversal t.

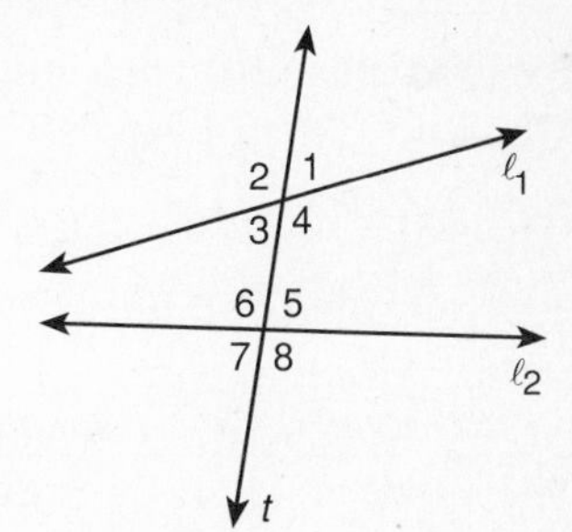

	Answers:
Angles 2 and 6	Corresponding angles
Angles 3 and 5	Alternate interior angles
Angles 3 and 6	Consecutive interior angles
Angles 2 and 8	Alternate exterior angles

b. Using the figure at the right, name pairs of congruent angles and supplementary angles. Note that $\ell_1 \parallel \ell_2$.

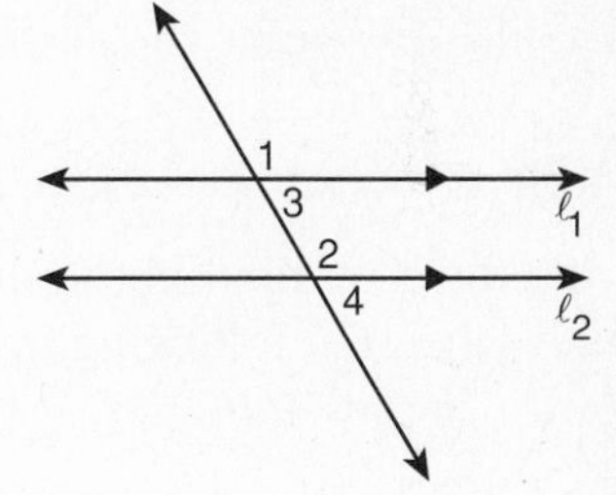

	Answers:
Congruent angles	$\angle 1$ and $\angle 2$, $\angle 3$ and $\angle 4$
Supplementary angles	$\angle 2$ and $\angle 3$, $\angle 2$ and $\angle 4$, $\angle 1$ and $\angle 3$, $\angle 1$ and $\angle 4$

Guidelines: If two parallel lines are cut by a transversal, then

- pairs of corresponding angles are congruent (postulate).
- pairs of alternate interior angles are congruent (theorem).
- pairs of alternate exterior angles are congruent (theorem).
- pairs of consecutive interior angles are supplementary (theorem).
- a transversal which is perpendicular to one of the parallel lines is perpendicular to the second (theorem).

EXERCISE

Use the figure at the right to find the measure of each labeled angle.

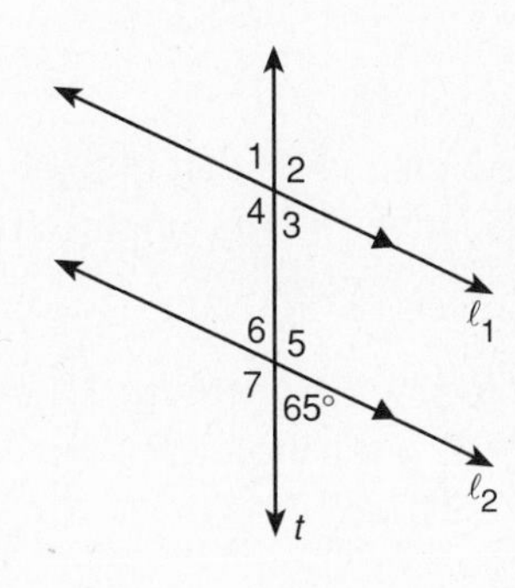

a. $m\angle 1$ b. $m\angle 2$ c. $m\angle 3$

d. $m\angle 4$ e. $m\angle 5$ f. $m\angle 6$

g. $m\angle 7$

Reteach
Chapter 3

Name ______________________________

What you should learn:

3.6	How to prove that two lines are parallel and use parallel lines in real life

Correlation to Pupil's Textbook:

Chapter Test (p. 157)
Exercises 19, 20

Examples *Proving that Two lines are Parallel and Using Parallel Lines*

a. Supply the reason for each statement in the two-column proof of the Alternate Exterior Angle Converse Theorem.

Given: $\angle 1 \cong \angle 2$
Prove: $\ell_1 \parallel \ell_2$
Proof:

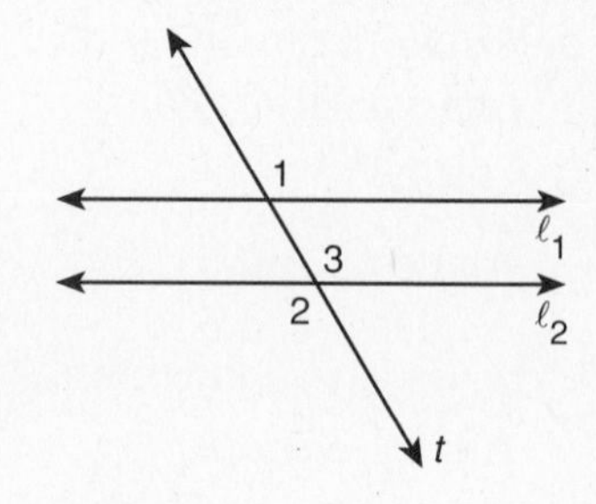

Statements	*Reasons*
1. $\angle 1 \cong \angle 2$	**1.** ? Given
2. $\angle 2 \cong \angle 3$	**2.** ? Vertical Angles Theorem
3. $\angle 1 \cong \angle 3$	**3.** ? Transitive Property of Congruence
4. $\ell_1 \parallel \ell_2$	**4.** ? Corresponding Angles Converse Postulate

b. Use the figure at the right to determine segments which are parallel when the following statements are true.

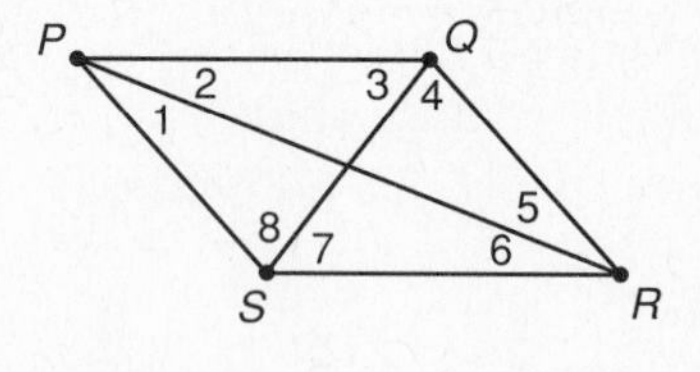

1. $\angle 3 \cong \angle 7$ *Answer:* $\overline{PQ} \parallel \overline{SR}$
2. $\angle 8 \cong \angle 4$ *Answer:* $\overline{PS} \parallel \overline{QR}$
3. $\angle QPS$, $\angle PSR$ are supplementary. *Answer:* $\overline{PQ} \parallel \overline{SR}$

Guidelines: Two lines cut by a transversal are parallel if

- corresponding angles are congruent or
- alternate interior angles are congruent or
- alternate exterior angles are congruent or
- consecutive interior angles are supplementary.

EXERCISES

1. Write the converse of Theorem 3.11–If two lines are cut by a transversal so that consecutive interior angles are supplementary, then the lines are parallel. Is the converse true?

2. Write the converse of Theorem 3.12–If two lines are cut by a transversal so that alternate exterior angles are congruent, then the lines are parallel. Is the converse true?

Reteach
Chapter 3

Name ______________________

What you should learn:

3.7	How to represent vectors as ordered pairs and add vectors

Correlation to Pupil's Textbook:

Chapter Test (p. 157)
Exercises 14–16

Examples *Representing Vectors as Ordered Pairs and Adding Vectors*

a. Let $R = (-5, 1)$ and $Q = (-2, -4)$. Find the ordered pair representation of $\overrightarrow{RQ}$ and the length of $\overrightarrow{RQ}$.

$\overrightarrow{RQ} = \langle x_2 - x_1, y_2 - y_1 \rangle = \langle 3, -5 \rangle$
$RQ = \sqrt{(x_2 - x_1)^2 + (y_2 - y_1)^2} = \sqrt{3^2 + (-5)^2} = \sqrt{34}$

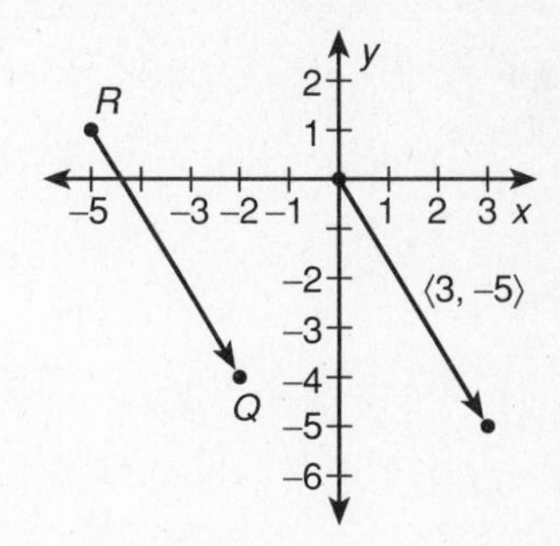

b. Let $\overrightarrow{v} = \langle 3, -2 \rangle$ and $\overrightarrow{u} = \langle 1, 2 \rangle$. Find the sum $\overrightarrow{v} + \overrightarrow{u}$.

$\overrightarrow{v} + \overrightarrow{u} = \langle a_1 + a_2, b_1 + b_2 \rangle = \langle 4, 0 \rangle$ which is a vector.

c. A small plane travels due south at a speed of 120 miles per hour. Its velocity is represented by $\overrightarrow{v} = \langle 0, -120 \rangle$. The plane encounters wind at 30 miles per hour from the west. The wind is represented by $\overrightarrow{w} = \langle 30, 0 \rangle$. Find the new speed of the plane.

$\overrightarrow{v} + \overrightarrow{w} = \langle 0 + 30, -120 + 0 \rangle = \langle 30, -120 \rangle$
New speed $= \sqrt{30^2 + (-120)^2} = \sqrt{15{,}300} \approx 124$ mph

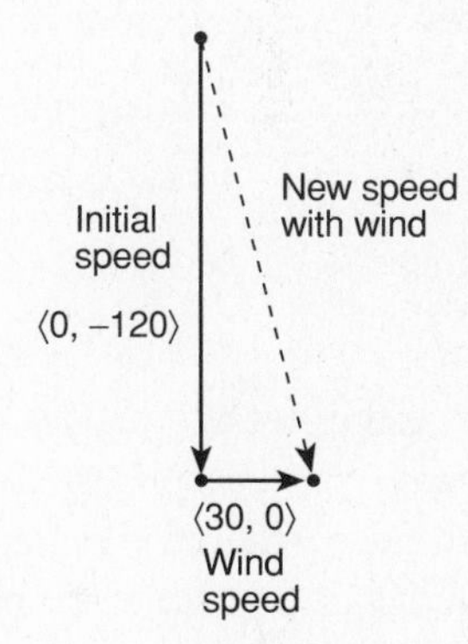

Guidelines:

- A *vector* is a quantity with direction and length.
- Two vectors are *parallel* if they have the same or opposite directions.
- Two vectors are *equal* if they have the same length and direction.
- One use of vectors is to represent velocity–rate and direction.

EXERCISES

In Exercises 1–4, find the sum of the vectors.

1. $\overrightarrow{v} = \langle 2, 1 \rangle, \overrightarrow{w} = \langle -2, 4 \rangle$
2. $\overrightarrow{u} = \langle 3, 6 \rangle, \overrightarrow{v} = \langle -3, -6 \rangle$
3. $\overrightarrow{z} = \langle 0, -4 \rangle, \overrightarrow{y} = \langle 5, 4 \rangle$
4. $\overrightarrow{s} = \langle -1, 3 \rangle, \overrightarrow{t} = \langle 6, 2 \rangle$

In Exercises 5 and 6, write the ordered pair representation of $\overrightarrow{AB}$ and $\overrightarrow{CD}$. Find the length of each vector and decide if they are equal.

5. $A = (-2, 1), B = (1, 1)$
   $C = (4, 3), D = (1, 3)$
6. $A = (-2, 2), B = (2, -1)$
   $C = (1, 5), D = (5, 2)$

Reteach
Chapter 3

Name ______________________

What you should learn:

3.7 How to determine if two vectors are perpendicular

Correlation to Pupil's Textbook:

Chapter Test (p. 157)
Exercises 17–18

Examples *Using the Dot Product of Two Vectors*

a. Let $\vec{v} = \langle 3, -2 \rangle$ and $\vec{u} = \langle 1, 2 \rangle$. Find the dot product $\vec{v} \cdot \vec{u}$.

$$\begin{aligned} \vec{v} \cdot \vec{u} &= a_1a_2 + b_1b_2 \\ &= 3(1) + (-2)(2) \\ &= 3 + (-4) \\ &= -1 \end{aligned}$$

The result is a real number.

b. Two vectors are perpendicular if and only if their dot product is zero. Show that the vectors $\vec{u} = \langle 6, -4 \rangle$ and $\vec{v} = \langle 4, 6 \rangle$ are perpendicular.

$$\begin{aligned} \vec{v} \cdot \vec{u} &= a_1a_2 + b_1b_2 \\ &= 4(6) + 6(-4) \\ &= 24 + (-24) \\ &= 0 \end{aligned}$$

Since the dot product is zero, the vectors are perpendicular.

c. Write the ordered pair representation of $\vec{v}$ and $\vec{u}$, then decide whether the vectors are perpendicular.

$\vec{v} = \langle x_2 - x_1, y_2 - y_1 \rangle = \langle 6 - 1, 1 - 1 \rangle = \langle 5, 0 \rangle$

$\vec{u} = \langle x_2 - x_1, y_2 - y_1 \rangle = \langle -2 - (-2), -3 - (-1) \rangle = \langle 0, -2 \rangle$

$\vec{v} \cdot \vec{u} = a_1a_2 + b_1b_2 = (5)0 + 0(-2) = 0$

Since the dot product is zero, the vectors are perpendicular.

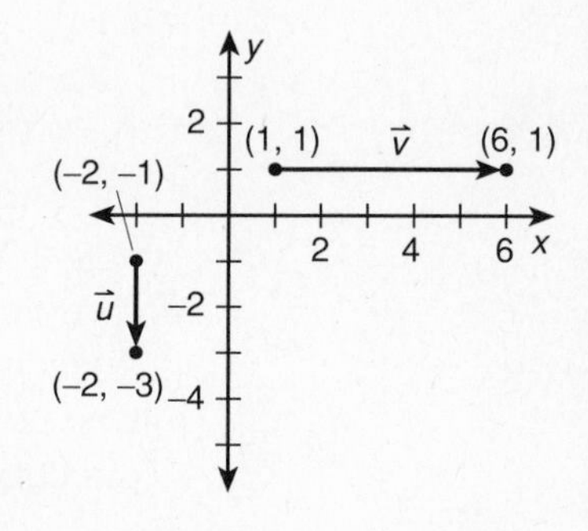

Guidelines:

- Two nonzero vectors are *perpendicular* if and only if their dot product is zero.
- The result of a *dot product* is a real number, not a vector.

EXERCISES

In Exercises 1–4, find the dot product of the vectors. Are the vectors perpendicular?

1. $\vec{v} = \langle 2, 1 \rangle, \vec{w} = \langle -2, 4 \rangle$
2. $\vec{u} = \langle 3, 6 \rangle, \vec{v} = \langle -3, -6 \rangle$
3. $\vec{z} = \langle 0, -4 \rangle, \vec{y} = \langle 5, 4 \rangle$
4. $\vec{s} = \langle -1, 3 \rangle, \vec{t} = \langle 6, 2 \rangle$

Reteach

Chapter 4

Name ______________________________

What you should learn:

4.1 How to identify congruent triangles and classify triangles by their sides and angles

Correlation to Pupil's Textbook:

Mid-Chapter Self-Test (p. 189)
Exercises 1–6, 11, 14–16

Chapter Test (p. 213)
Exercises 1–3, 17–20

Examples *Identifying Congruent Triangles and Classifying Triangles*

a. In the figure at the right, $\triangle RST \cong \triangle MNO$. This means that corresponding sides are congruent and corresponding angles are congruent.

Corresponding angles: $\angle R \cong \angle M$, $\angle S \cong \angle N$, $\angle T \cong \angle O$

Corresponding sides: $\overline{RS} \cong \overline{MN}$, $\overline{ST} \cong \overline{NO}$, $\overline{TR} \cong \overline{OM}$

b. The triangles in the figure at the right can be classified by their sides as follows.

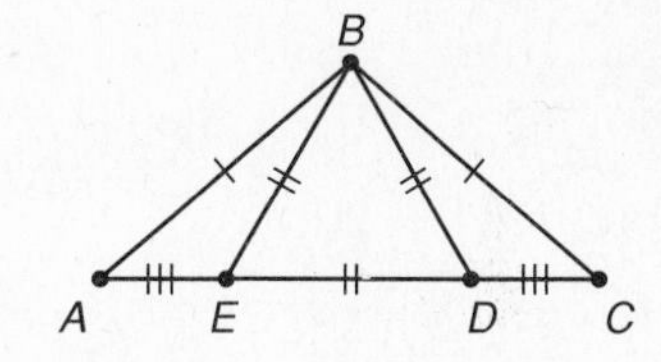

Isosceles (at least two congruent sides): $\triangle ABC$, $\triangle EBD$

Equilateral (three congruent sides): $\triangle EBD$

Scalene (no congruent sides): $\triangle ABE$, $\triangle CBD$, $\triangle ABD$, $\triangle CBE$

c. The triangles in the figures at the right can be classified by their angles as follows.

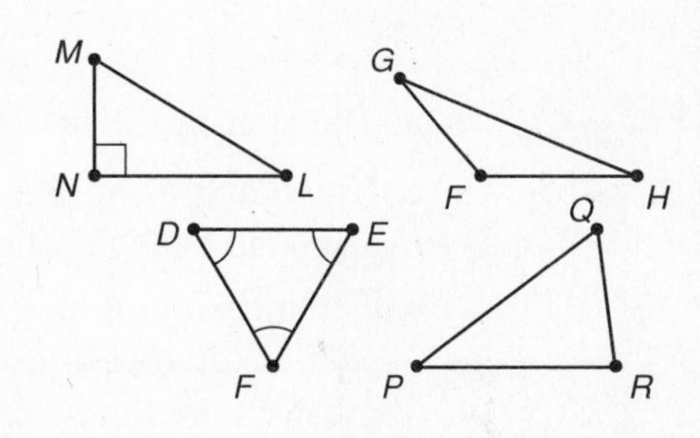

Acute (three acute angles): $\triangle DEF$, $\triangle PQR$

Right (exactly one right angle): $\triangle NML$

Obtuse (exactly one obtuse angle): $\triangle FGH$

Equiangular (three congruent acute angles): $\triangle DEF$

Guidelines:

- In $\triangle ABC$, points A, B, and C are *vertices*.
- In $\triangle ABC$, $\overline{AB}$, $\overline{BC}$, and $\overline{CA}$ are *sides*.
- *Adjacent sides* of a triangle share a common vertex.
- In a right triangle, the sides adjacent to the right angle are *legs* and the side opposite the right angle is the *hypotenuse*.
- Congruence of triangles is *reflexive, symmetric,* and *transitive*.

EXERCISES

1. If $\triangle LMN \cong \triangle PQR$, then $m\angle N = \boxed{?}$.

2. If $\triangle LMN \cong \triangle PQR$, then $\overline{MN} \cong \boxed{?}$.

In Exercises 3–6, use the figure at the right to complete the statement.

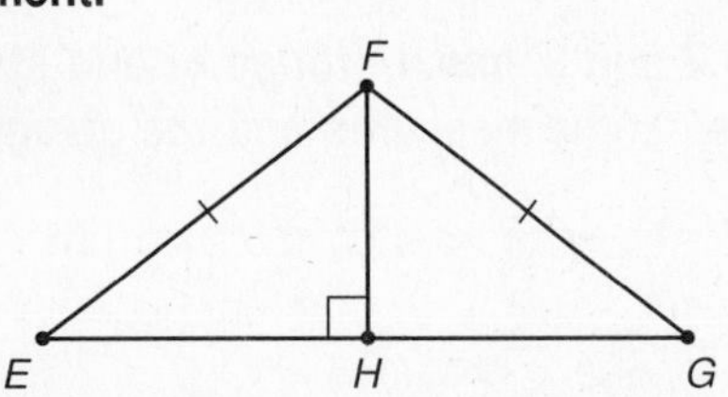

3. The hypotenuse of $\triangle EFH$ is $\boxed{?}$.

4. The base of the isosceles triangle, $\triangle EFG$, is $\boxed{?}$.

5. The legs of $\triangle FHG$ are $\boxed{?}$ and $\boxed{?}$.

6. If $\triangle EFH \cong \triangle GFH$, then $\triangle GFH \cong \boxed{?}$.

Reteach
Chapter 4

Name ____________________

What you should learn:

4.2	How to measure the angles of a triangle and use angle measures in real-life problems

Correlation to Pupil's Textbook:

Mid-Chapter Self-Test (p. 189) Exercises 7–10

Chapter Test (p. 213) Exercises 5–10

Examples *Finding Angle Measures and Using Angle Measures in Real Life*

a. In the triangle at the right, determine $m\angle 1$ and $m\angle 2$.

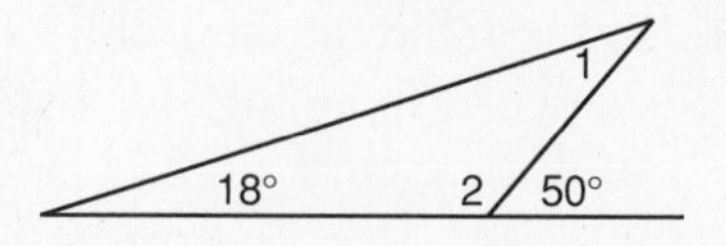

$m\angle 2 = 180° - 50° = 130°$ by the Linear Pair Postulate

$m\angle 1 = 180° - (18 + 130) = 32°$ by the Triangle Sum Theorem

b. You are building a brick walkway with a corner turn, as shown at the right. If $m\angle 2 = 70°$, what is $m\angle 1$?

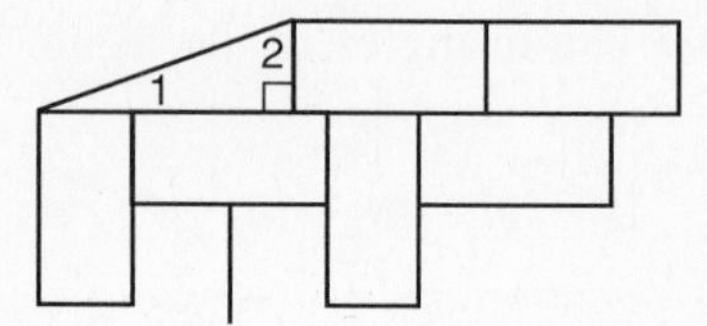

By Theorem 4.4, the acute angles of a right triangle are complementary; therefore, $m\angle 1 = 20°$.

Guidelines:

- The sum of the interior angles of a triangle is 180°.
- If two angles of one triangle are congruent to two angles of a second triangle, then the third angles are also congruent.
- The measure of an exterior angle of a triangle is equal to the sum of the measures of the two remote interior angles, and is greater than the measure of either of the two remote interior angles.

EXERCISES

In Exercises 1–6, find the measure of each labeled angle.

1. $m\angle 1$
2. $m\angle 2$
3. $m\angle 3$
4. $m\angle 4$
5. $m\angle 5$
6. $m\angle 6$

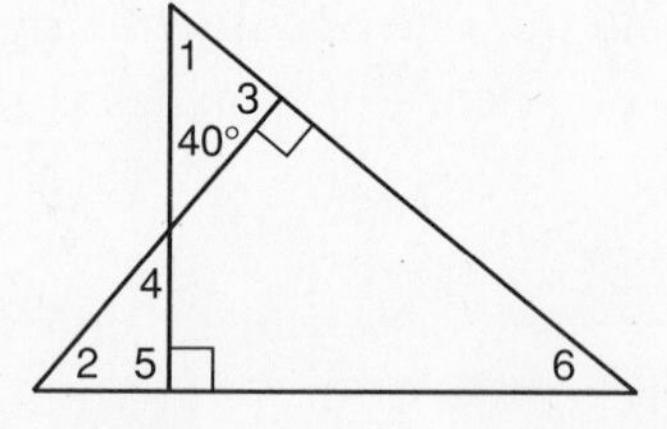

In Exercises 7 and 8, use a protractor and straightedge to draw $\triangle ABC$ with the given angle measures and the given length of the included side.

7. $m\angle A = 50°, m\angle B = 35°, AB = 3$ cm

8. $m\angle A = 30°, m\angle B = 60°, AB = 2$ in.

Reteach
Chapter 4

Name ______________________________

What you should learn:

4.3	How to use SSS or SAS Congruence Postulates to verify congruent triangles and solve real-life problems

Correlation to Pupil's Textbook:

Mid-Chapter Self-Test (p. 189)
Exercises 19, 20

Chapter Test (p. 213)
Exercises 11–14

Examples *Using SSS and SAS Congruence Postulates*

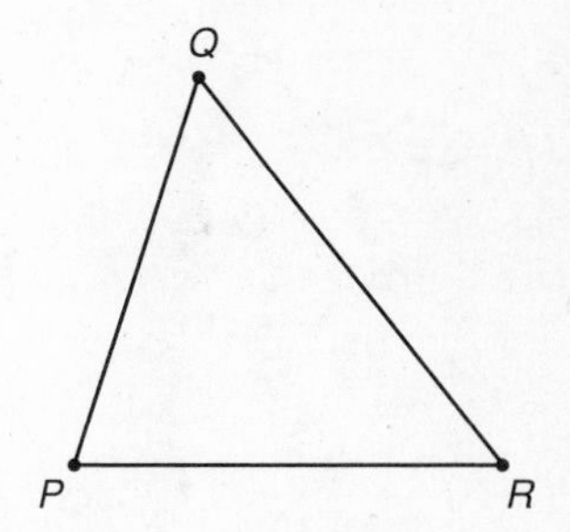

a. Use the figure at the right to answer the following questions.

1. Between what two sides is $\angle P$ included?
2. What angle is included between sides $\overline{RQ}$ and $\overline{QP}$?
3. Name two angles which are not included between sides $\overline{PR}$ and $\overline{RQ}$.

Answers: 1. $\overline{QP}$ and $\overline{PR}$ 2. $\angle Q$ 3. $\angle Q$ and $\angle P$

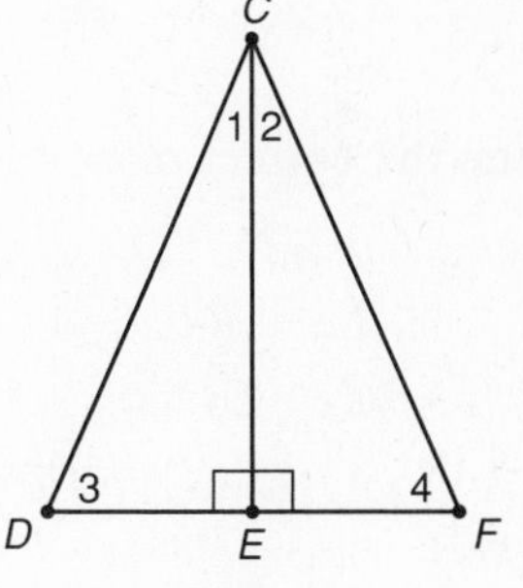

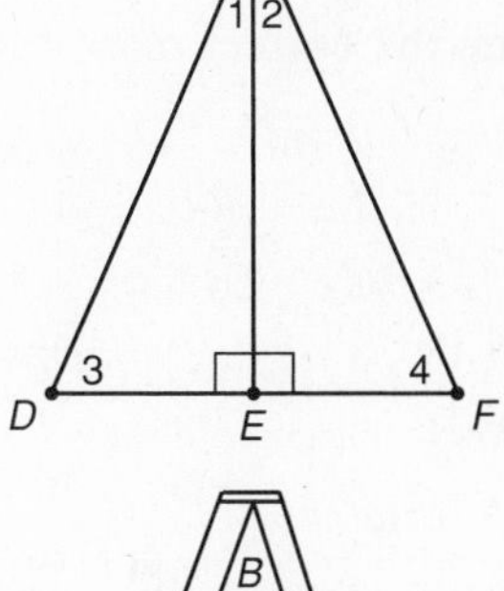

b. Using the figure at the right and the given information, state the postulate which can be used to verify that $\triangle CDE \cong \triangle CFE$.

	Answers:
1. Given: $\overline{CD} \cong \overline{CF}$, $\overline{DE} \cong \overline{FE}$	SSS Congruence Postulate
2. Given: $\overline{CD} \cong \overline{CF}$, $\angle 1 \cong \angle 2$	SAS Congruence Postulate
3. Given: $\overline{DE} \cong \overline{FE}$	SAS Congruence Postulate

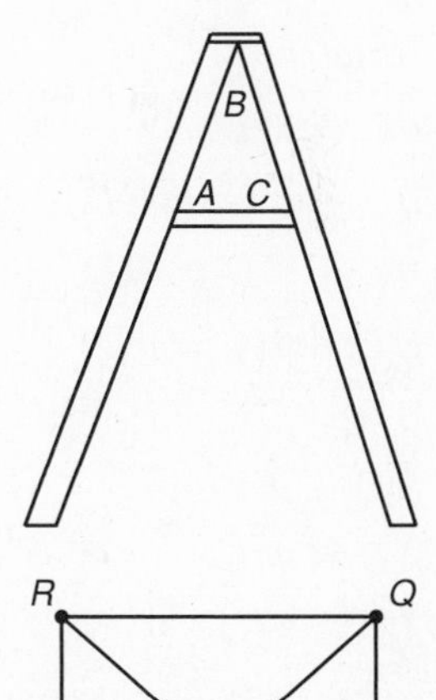

c. A stepladder has a hinged side brace for safety support. Explain why the brace forms a rigid framework.

The brace forms a triangle ABC with fixed segment lengths.

d. Write a proof that $\triangle RPQ \cong \triangle QSR$.

Given: $\overline{RP} \cong \overline{QS}$
$\overline{RS} \cong \overline{QP}$

Proof:

Statements	*Reasons*
1. $\overline{RP} \cong \overline{QS}$	**1.** Given
2. $\overline{RS} \cong \overline{QP}$	**2.** Given
3. $\overline{RQ} \cong \overline{QR}$	**3.** Reflexive Prop. of Congruence
4. $\triangle RPQ \cong \triangle QSR$	**4.** SSS Congruence Postulate

Reteach
Chapter 4

Name ______________________

Guidelines:

- If three sides of one triangle are congruent to three sides of a second triangle, then the two triangles are congruent (SSS Congruence Postulate).
- If two sides and the included angle of one triangle are congruent to two sides and the included angle of a second triangle, then the two triangles are congruent (SAS Congruence Postulate).

EXERCISES

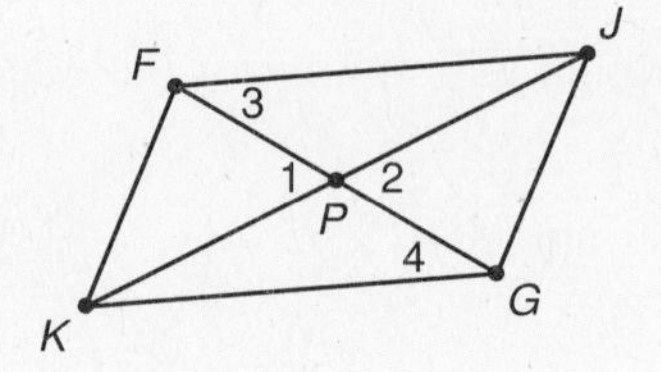

1. Write a proof that $\triangle FKP \cong \triangle GJP$, given that $\overline{FG}$ and $\overline{JK}$ bisect each other at P.

2. Write a proof that $\triangle FJG \cong \triangle GKF$, given that $\overline{FJ} \parallel \overline{KG}$ and $\overline{FJ} \cong \overline{KG}$.

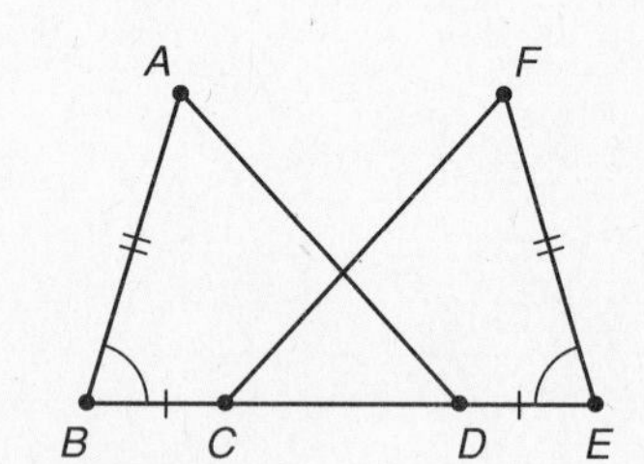

3. Match the correct reason for each step in the proof.

Given: $\overline{AB} \cong \overline{FE}$

$\angle B \cong \angle E$

$BC = ED$

Prove: $\triangle ABD \cong \triangle FEC$

Proof:

Statements	*Reasons*
1. $BC = ED$	?
2. $BC + CD = ED + DC$	?
3. $BC + CD = BD$	?
4. $ED + DC = EC$	?
5. $BD = EC$	?
6. $\overline{BD} \cong \overline{EC}$	?
7. $\overline{AB} \cong \overline{FE}$	?
8. $\angle B \cong \angle E$	?
9. $\triangle ABD \cong \triangle FEC$	?

a. Def. of Congruence
b. Given
c. SAS Congruence Postulate
d. Segment Addition Postulate
e. Addition Prop. of Equality
f. Given
g. Substitution Prop. of Equality
h. Segment Addition Postulate
i. Given

Reteach
Chapter 4

Name ______________________________

What you should learn:

4.4	How to use ASA Congruence Postulate or AAS Congruence Theorem and to solve real-life problems

Correlation to Pupil's Textbook:

Mid-Chapter Self-Test (p. 189)
Exercises 11, 13, 15–18

Chapter Test (p. 213)
Exercises 4, 11–14

Examples *Using ASA Congruence Postulate and AAS Congruence Theorem*

a. State the postulate or theorem you would use with the given information to prove $\triangle OMN \cong \triangle DEF$.

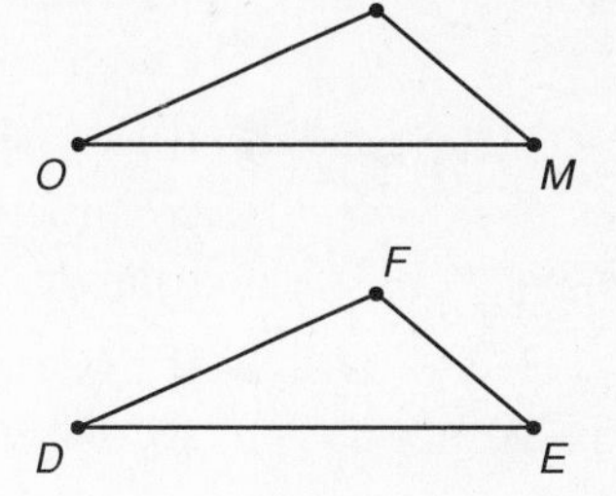

	Answers:
1. $\angle N \cong \angle F, \angle M \cong \angle E, \overline{ON} \cong \overline{DF}$	AAS Congruence Theorem
2. $\angle O \cong \angle D, \overline{OM} \cong \overline{DE}, \angle M \cong \angle E$	ASA Congruence Postulate
3. $\angle N \cong \angle F, \angle O \cong \angle D, \overline{OM} \cong \overline{DE}$	AAS Congruence Theorem

b. Write a paragraph proof, using the given figure.

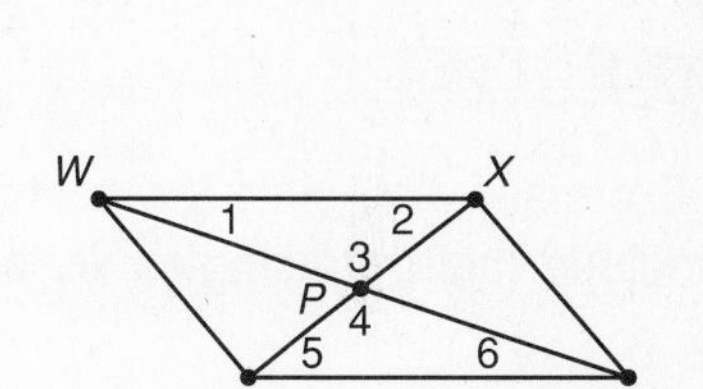

Given: $\overline{WX} \parallel \overline{YZ}, \overline{WP} \cong \overline{YP}$

Prove: $\triangle WPX \cong \triangle YPZ$

Proof:

- Since $\overline{WX} \parallel \overline{YZ}$, $\angle 1 \cong \angle 6$ by the Alternate Interior Angles Theorem.
- Vertical angles $\angle 3$ and $\angle 4$ are congruent.
- Given congruent sides $\overline{WP} \cong \overline{YP}$ are included sides.
- By the ASA Congruence Postulate, $\triangle WPX \cong \triangle YPZ$.

c. You are playing a video game in which you are the pilot of a jet aircraft. The game directs you to fly 400 miles due north. Next you are instructed to fly 500 miles at a bearing of S50°E, or 50° to the east of south. Do you have enough information to determine the aircraft's final destination?

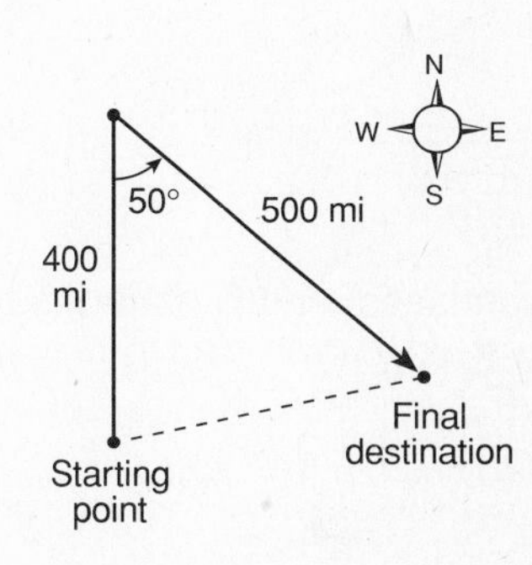

Yes, you can interpret the given information in the following way.

Because you are given two sides and the included angle of the triangle you can determine the location of the third vertex.

Reteach
Chapter 4

Name ____________________

Guidelines:

- If two angles and the included side of one triangle are congruent to two angles and the included side of a second triangle, then the triangles are congruent (ASA Congruence Postulate).
- If two angles and a nonincluded side of one triangle are congruent to two angles and the corresponding nonincluded side of a second triangle, then the triangles are congruent (AAS Congruence Theorem).
- If two sides and a nonincluded angle of one triangle are congruent to two sides and a nonincluded angle of a second triangle, then the triangles are *not* necessarily congruent.
- If three angles of one triangle are congruent to three angles of a second triangle, then the triangles are *not* necessarily congruent.

EXERCISES

In Exercises 1–4, state the postulate or theorem that can be used to conclude that the triangles are congruent.

1.

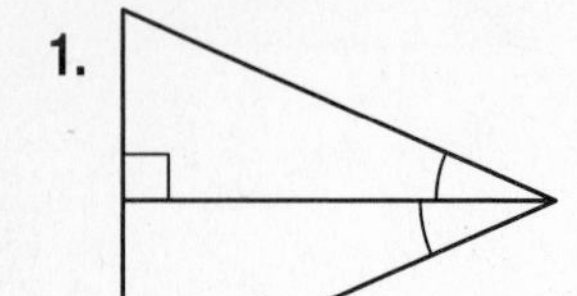

2.

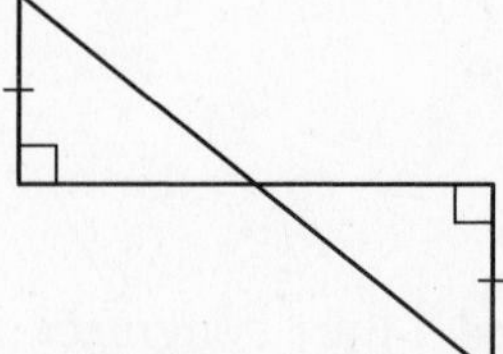

3.

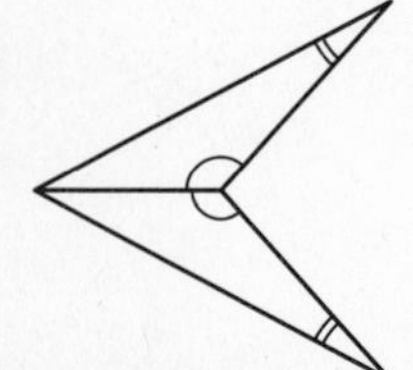

4.

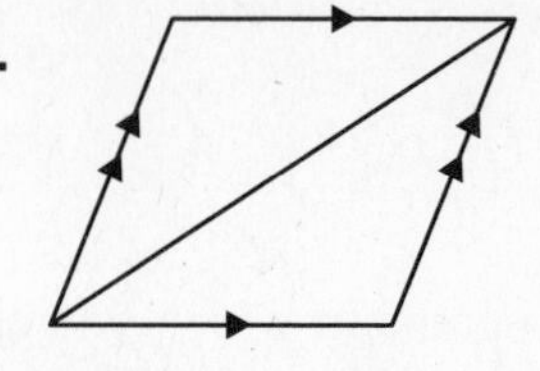

In Exercises 5 and 6, state the third congruence that must be given to prove that $\triangle ABC \cong \triangle DEF$, using the indicated method.

5. *Given:* $\angle A \cong \angle D$

$\angle B \cong \angle E$

Method: AAS

6. *Given:* $\angle A \cong \angle D$

$\angle B \cong \angle E$

Method: ASA

Reteach

Chapter 4

Name ______________________________

What you should learn:

4.5	How to plan a proof. How to prove constructions are valid

Correlation to Pupil's Textbook:

Chapter Test (p. 213)

Exercises 3, 15, 16

Examples — *Planning a Proof and Proving Constructions*

a. Plan a proof by marking the diagram with congruence marks and using the definition that two triangles are congruent if and only if their corresponding parts are congruent.

Given: $\angle 1 \cong \angle 2$, $\angle 3 \cong \angle 4$

Prove: $\overline{BC} \cong \overline{DC}$

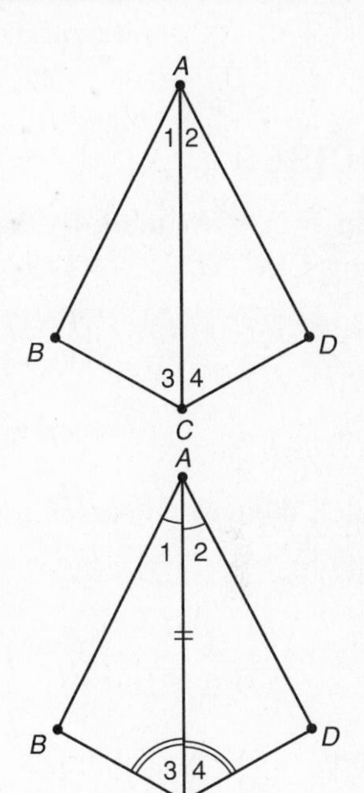

Planning a proof:

Mark the diagram with a given information. Note that $\overline{AC} \cong \overline{AC}$ by the Reflexive Property of Congruence. This implies that $\triangle BAC \cong \triangle DAC$ by the ASA Congruence Postulate. Since the triangles are congruent, you can conclude that $\overline{BC} \cong \overline{DC}$ because CPCTC.

b. Use a straightedge and compass to bisect a right angle. Then write a paragraph proof to verify the results.

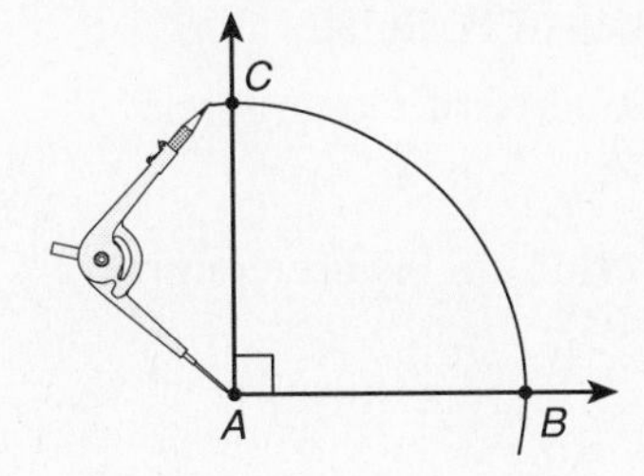

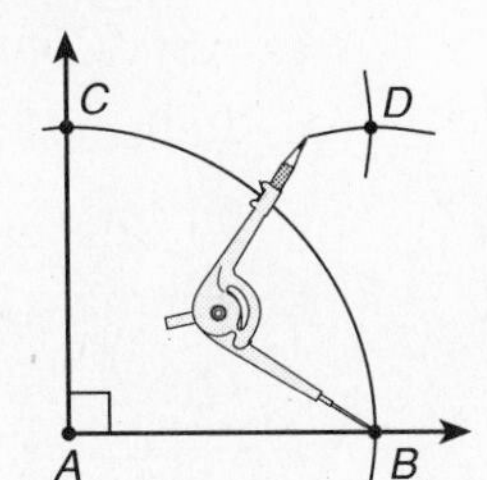

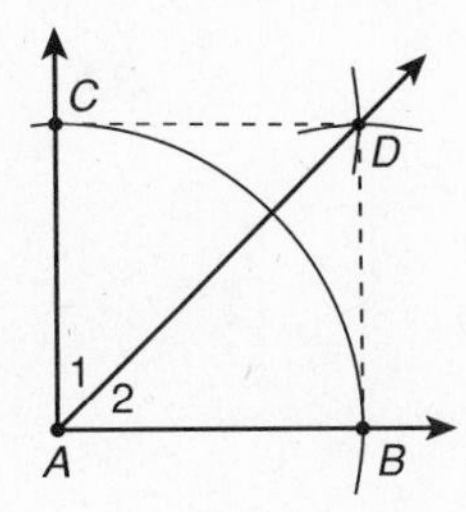

- By construction, $\overline{AB} \cong \overline{AC}$ and $\overline{BD} \cong \overline{CD}$.
- Since $\overline{AD} \cong \overline{AD}$, by the Reflexive Prop. of Congruence, $\triangle ACD \cong \triangle ABD$ by the SSS Congruence Postulate.
- By CPCTC, it follows that $\angle 1 \cong \angle 2$.
- By definition, $\overrightarrow{AD}$ bisects right angle A.

Reteach
Chapter 4

Name ______________________________

Guidelines:

- You can use congruent triangles and the definition Corresponding Parts of Congruent Triangles are Congruent (CPCTC) for planning proofs.
- You can use congruent triangles to prove that basic constructions are valid.

EXERCISES

1. Plan a proof by marking the diagram with congruence marks and using CPCTC.

Given: $\overline{PQ} \parallel \overline{RS}$, $\overline{PQ} \cong \overline{RS}$
Prove: $\overline{PS} \parallel \overline{QR}$

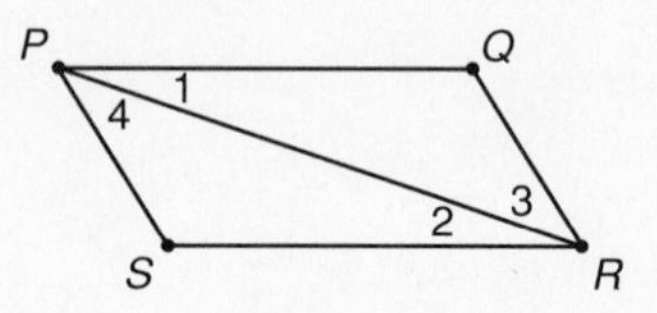

2. Match the correct reason for each step in the proof.

Given: $\overline{AB} \parallel \overline{CD}$
$\overline{BC} \parallel \overline{DA}$
Prove: $\overline{AD} \cong \overline{CB}$

Proof:

Statements	*Reasons*
1. $\overline{AB} \parallel \overline{CD}$	?
2. $\angle BAC \cong \angle DCA$	?
3. $\overline{BC} \parallel \overline{DA}$	?
4. $\angle BCA \cong \angle DAC$	?
5. $\overline{AC} \cong \overline{AC}$	?
6. $\triangle BAC \cong \triangle DCA$	?
7. $\overline{AD} \cong \overline{CB}$	?

a. Alternate interior angles are congruent.
b. CPCTC
c. ASA Congruence Postulate
d. Reflexive Property of Congruence
e. Given
f. Alternate interior angles are congruent.
g. Given

Reteach
Chapter 4

Name ____________________

What you should learn:

4.6	How to use properties of isosceles triangles and right triangles to solve problems

Correlation to Pupil's Textbook:

Chapter Test (p. 213)
Exercise 15

Examples *Using Properties of Isosceles Triangles and Right Triangles*

a. Use the triangles at the right to identify the following:

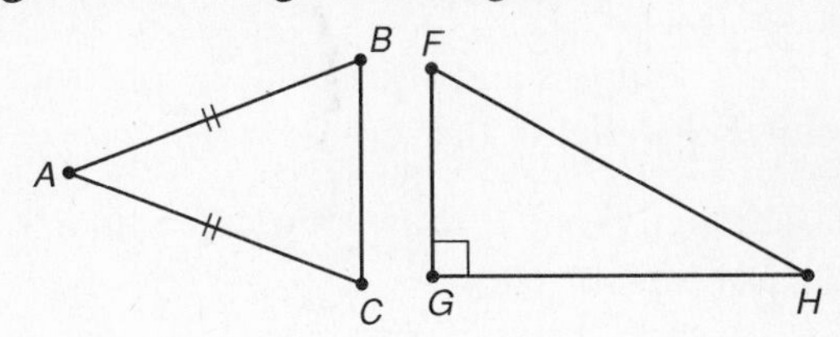

	Answers:
1. Base of $\triangle ABC$	$\overline{BC}$
2. Legs of $\triangle FGH$	$\overline{FG}, \overline{GH}$
3. Base angles of $\triangle ABC$	$\angle B, \angle C$
4. Hypotenuse of $\triangle FGH$	$\overline{FH}$
5. Congruent angles	$\angle B \cong \angle C$

b. Using the figure at the right, write a flow proof.

Given: $\overline{HJ} \cong \overline{LJ}, \overline{HI} \cong \overline{LK}$; $\angle HIJ$ is a right angle. $\angle LKJ$ is a right angle.

Prove: $\triangle HIJ \cong \triangle LKJ$

Proof:

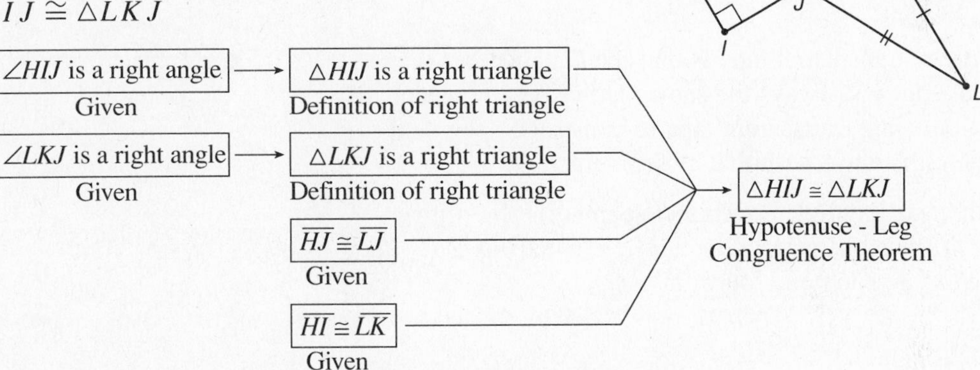

Guidelines:

- If two sides of a triangle are congruent, then the angles opposite them are congruent (Base Angles Theorem).
- If two angles of a triangle are congruent, then the sides opposite them are congruent.
- If a triangle is equilateral, then it is also equiangular.
- If a triangle is equiangular, then it is also equilateral.
- If the hypotenuse and a leg of a right triangle are congruent to the hypotenuse and leg of a second right triangle, then the two triangles are congruent (Hypotenuse-Leg Congruence Theorem).

EXERCISE

Write a proof for the corollary to Theorem 4.8—If a triangle is equilateral, then it is also equiangular.

Given: $\overline{AB} \cong \overline{BC} \cong \overline{CA}$ *Prove:* $\triangle ABC$ is equiangular.

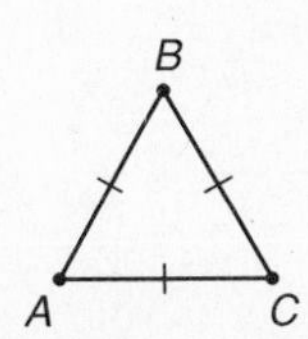

Reteach
Chapter 4

Name ______________________________

What you should learn:

4.7	Using congruent triangles to prove both straightedge and compass constructions and nonstandard constructions

Correlation to Pupil's Textbook:

Chapter Test (p. 213)
Exercise 3

Examples *Proving Straightedge/Compass and Nonstandard Constructions*

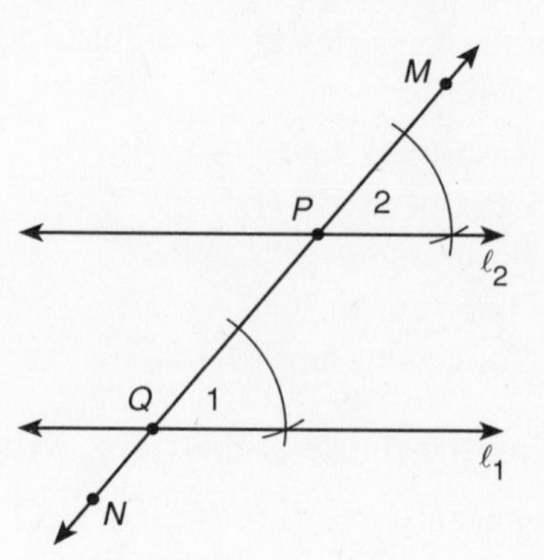

a. Using a compass and straightedge, construct a line through point P parallel to line ℓ_1.

1. Through point P draw an oblique line $\overleftrightarrow{MN}$ intersecting line ℓ_1 at a point Q.
2. With vertex P and $\overleftrightarrow{MN}$ as a transversal, construct $\angle 2$ congruent to $\angle 1$.
3. Since $\angle 1$ and $\angle 2$ are congruent corresponding angles, $\ell_1 \parallel \ell_2$.

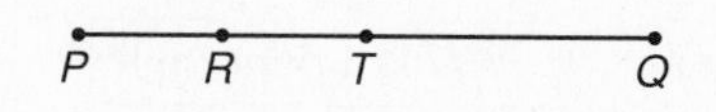

b. A property of transparent tape is that the distance between its parallel edges is always the same. Describe a nonstandard construction using transparent tape to construct a segment whose length is one-fourth that of segment $\overline{PQ}$.

Follow the text Example 2 to bisect segment $\overline{PQ}$, so that PT is half of PQ. Then bisect $\overline{PT}$ so that PR is half of PT. Then PR and RT are each one fourth of PQ.

Guidelines:

- The *distance* between parallel lines is the length of a perpendicular segment between the lines.
- When two parallel lines intersect two other parallel lines and the distance between the first two lines is equal to the distance between the second two lines, then their intersection forms four congruent segments.

EXERCISE

Write the reason for each step in the proof of the construction in Example a above.

Given: $\overline{QR} \cong \overline{PS}$, $\overline{QT} \cong \overline{PV}$, $\overline{RT} \cong \overline{SV}$

Prove: $\ell_2 \parallel \ell_1$

Statements	*Reasons*
1. $\overline{QR} \cong \overline{PS}$	**1.** ?
2. $\overline{QT} \cong \overline{PV}$	**2.** ?
3. $\overline{RT} \cong \overline{SV}$	**3.** ?
4. $\triangle QRT \cong \triangle PSV$	**4.** ?
5. $\angle 1 \cong \angle 2$	**5.** ?
6. $\ell_2 \parallel \ell_1$	**6.** ?

Reteach

Chapter 5

Name ______________________________

What you should learn:

5.1	How to use perpendicular bisectors and angle bisectors

Correlation to Pupil's Textbook:

Mid-Chapter Self-Test (p. 239)
Exercises 2, 3, 6, 8, 10, 11

Chapter Test (p. 263)
Exercises 1, 3

Examples *Using Perpendicular Bisectors and Angle Bisectors*

a. In the figure at the right, $\overrightarrow{XZ}$ is the angle bisector of $\angle WXY$.

1. Name two congruent angles. *Answer:* $\angle WXZ \cong \angle YXZ$
2. Name two congruent segments. *Answer:* $\overline{ZW} \cong \overline{ZY}$

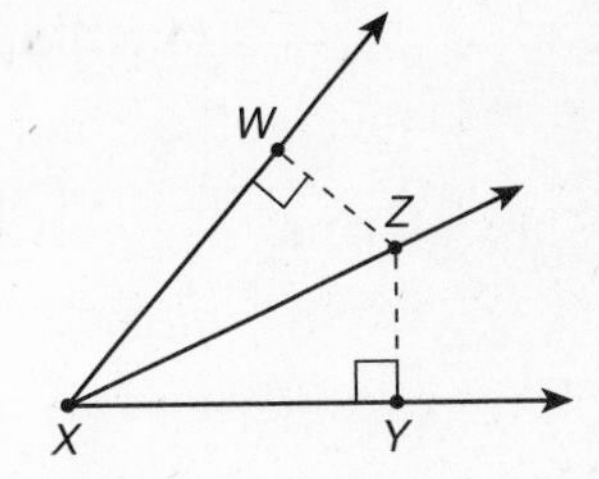

b. In the figure at the right $\overline{AD} \cong \overline{CD}$. What does this tell you about $\angle ABD$ and $\angle CBD$?

Answer: Because point D lies in the interior of $\angle ABC$ and is equidistant from the sides $\overrightarrow{BA}$ and $\overrightarrow{BC}$, we know by the Angle Bisector Converse that D lies on the angle bisector of $\angle ABC$.

Because $\overrightarrow{BD}$ is the angle bisector of $\angle ABC$, $\angle ABD \cong \angle CBD$.

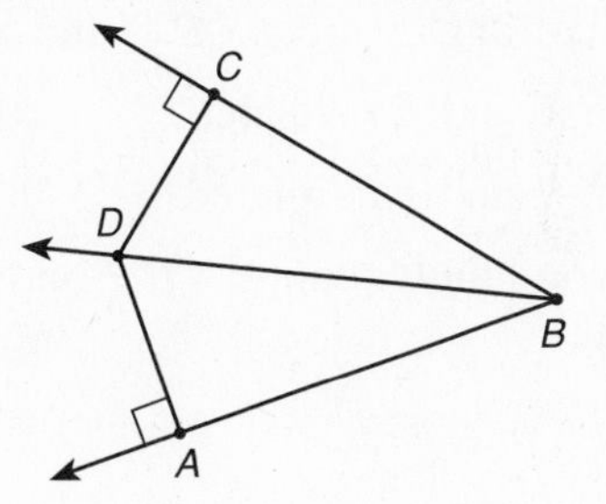

c. Using the figure at the right, sketch the perpendicular bisector of $\overline{XY}$.

You can find the midpoint of $\overline{XY}$ by using the Midpoint Formula

$$\left(\frac{x_1 + x_2}{2}, \frac{y_1 + y_2}{2}\right)$$

for the points $X(2, 8)$ and $Y(10, 10)$.
The coordinates of the midpoint, M, are

$$\left(\frac{2 + 10}{2}, \frac{8 + 10}{2}\right) = (6, 9).$$

You can find the slope of $\overline{XY}$ by using the slope formula.

$$m = \frac{y_2 - y_1}{x_2 - x_1} = \frac{10 - 8}{10 - 2} = \frac{2}{8} = \frac{1}{4}$$

Since a line perpendicular to $\overline{XY}$ has slope -4, you can sketch a line passing through point $M(6, 9)$ with slope -4.

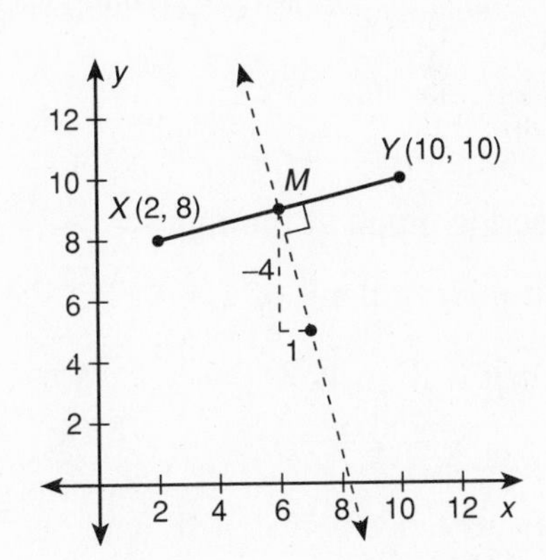

Reteach
Chapter 5

Name ______________________________

Guidelines:

- If a point is on the perpendicular bisector of a segment, then it is equidistant from the endpoints of the segment.
- If a point is equidistant from the endpoints of a segment, then it lies on the perpendicular bisector of the segment.
- If a point is on the bisector of an angle, then it is equidistant from the two sides of the angle.
- If a point is in the interior of an angle and equidistant from the sides of an angle, then it lies on the bisector of the angle.

EXERCISES

In 1–4, use the figure at the right where $\overrightarrow{PQ}$ is a perpendicular bisector of $\overline{XY}$.

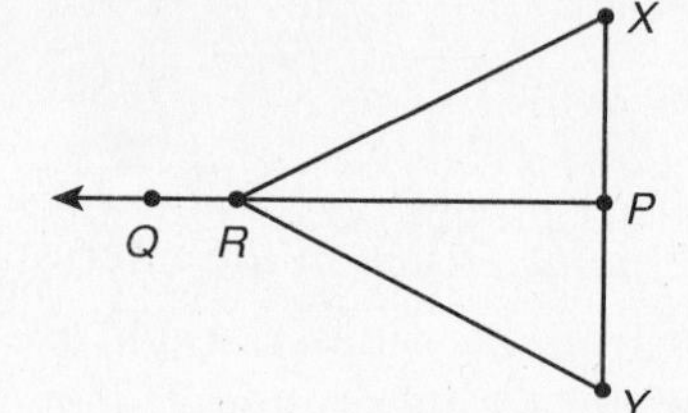

1. What type of angles are $\angle XPQ$ and $\angle YPQ$?
2. How are $\overline{RX}$ and $\overline{RY}$ related?
3. How are $\overline{XP}$ and $\overline{YP}$ related?
4. What do you know about $\angle XRP$ and $\angle YRP$?

In 5 and 6, use the information given in Example c on page 39.

5. Find the equation of the perpendicular bisector of $\overline{XY}$.
6. If point Q is on the perpendicular bisector of $\overline{XY}$, how are XQ and YQ related?

In 7–12, use the graph at the right.

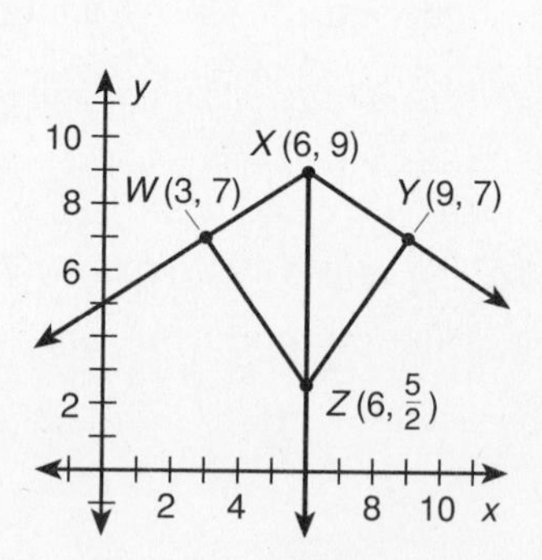

7. Is Z an interior point of $\angle WXY$?
8. Show that $\overrightarrow{XW} \perp \overline{WZ}$.
9. Show that $\overrightarrow{XY} \perp \overline{YZ}$.
10. Find the distance between W and Z.
11. Find the distance between Y and Z.
12. Explain why Z lies on the angle bisector of $\angle WXY$.

Reteach
Chapter 5

Name ______________________________

What you should learn:

5.2 How to identify special segments in a triangle and use concurrency properties

Correlation to Pupil's Textbook:

Mid-Chapter Self-Test (p. 239) Exercises 2, 4–11, 13, 15–19

Chapter Test (p. 263) Exercises 3–8

Examples *Identifying Special Segments and Using Concurrency Properties*

a. Use the figures at the right to identify each of the following special segments.

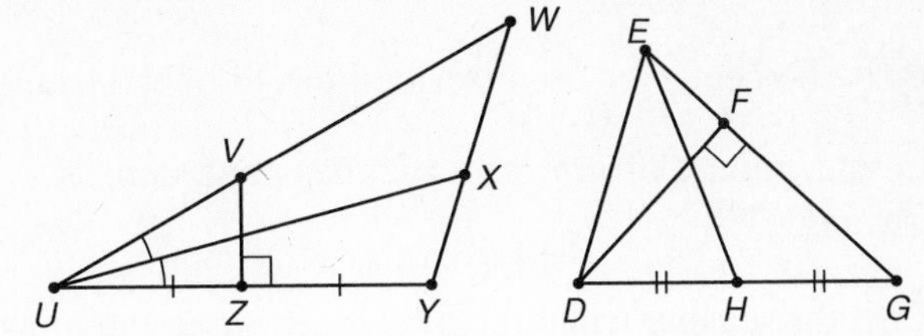

1. Altitude
2. Angle bisector
3. Perpendicular bisector
4. Median

Answers: 1. $\overline{DF}$ 2. $\overline{UX}$ 3. $\overline{VZ}$ 4. $\overline{EH}$

b. Match the following common points with their special names.

1. Intersection of the medians — a. incenter
2. Intersection of the altitudes — b. centroid
3. Intersection of the angle bisectors — c. orthocenter
4. Intersection of the perpendicular bisectors — d. circumcenter

Answers: 1. b 2. c 3. a 4. d

c. An altitude of a triangle may lie inside or outside the triangle. Identify all three altitudes of $\triangle ABC$ shown at the right.

The altitudes are $\overline{AD}$, $\overline{CE}$, and $\overline{BF}$.

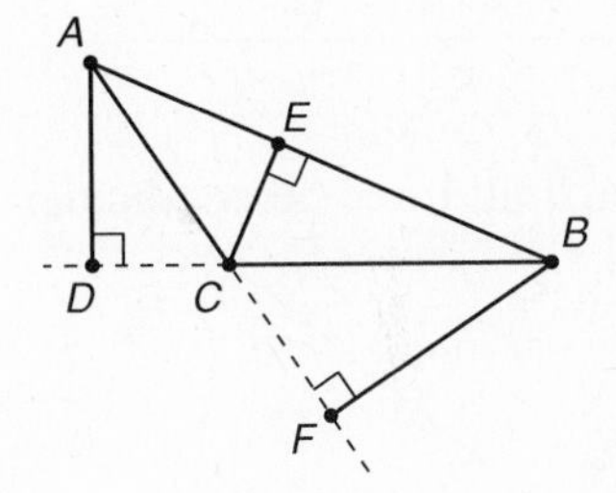

Guidelines:

- Angle bisectors, medians, and altitudes always have a vertex as one endpoint; perpendicular bisectors may or may not have a vertex as an endpoint.
- An altitude may lie inside or outside the triangle (An altitude lies on a right triangle.), thus the location of the orthocenter varies.
- The incenter is equidistant from the three sides of the triangle.
- The centroid is two-thirds of the distance from each vertex to the midpoint of the opposite side.
- The circumcenter is equidistant from the three vertices of the triangle.

Name ________________________________

EXERCISES

In 1–3, complete the statement.

1. The [?] is the center of the circumscribed circle that passes through the vertices of the triangle.
2. The [?] is the center of the inscribed circle of the triangle.
3. A triangular model of uniform thickness and density will balance at the [?] of the triangle.
4. Draw a right scalene triangle. Sketch all three medians. Label your sketch to demonstrate your geometric vocabulary.
5. Draw an obtuse isosceles triangle. Sketch all three angle bisectors. Label your sketch to demonstrate your geometric vocabulary.
6. Sketch all three altitudes for the triangle.

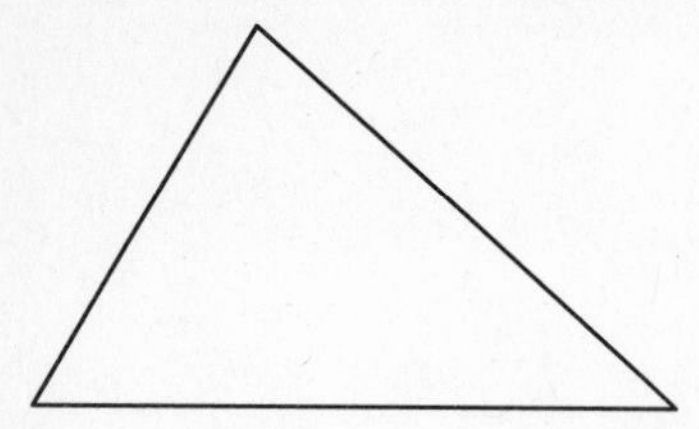

7. Sketch all three perpendicular bisectors for the triangle.

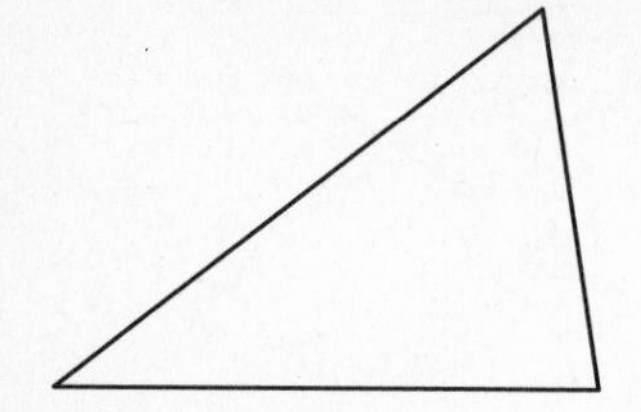

Reteach
Chapter 5

Name ______________________________

What you should learn:

5.3 How to identify and construct midsegments of a triangle and solve problems

Correlation to Pupil's Textbook:

Mid-Chapter Self-Test (p. 239) Exercises 1, 12, 14, 20

Chapter Test (p. 263) Exercises 2, 9–12

Examples *Constructing Midsegments and Using Properties of Midsegments*

a. Show that the midsegment $\overline{HF}$ is parallel to side $\overline{DE}$ and half its length.

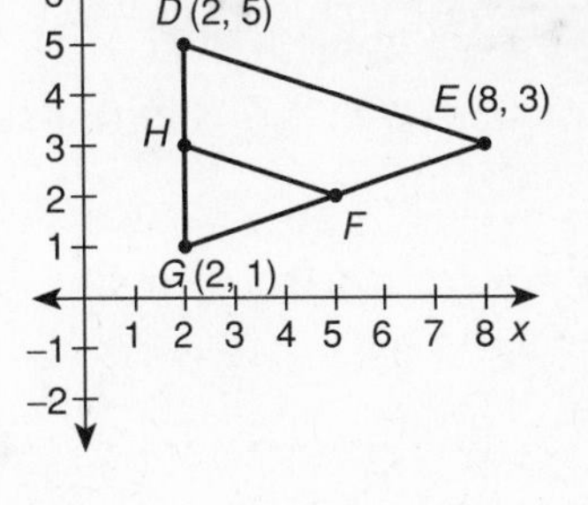

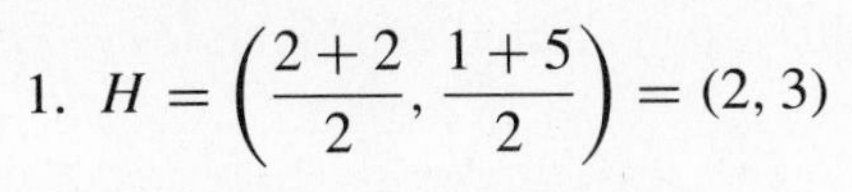

1. $H = \left(\frac{2+2}{2}, \frac{1+5}{2}\right) = (2, 3)$

   $F = \left(\frac{2+8}{2}, \frac{1+3}{2}\right) = (5, 2)$

   The slope of $\overline{HF} = -\frac{1}{3}$ and the slope of $\overline{DE} = -\frac{1}{3}$, thus $\overline{HF} \parallel \overline{DE}$.

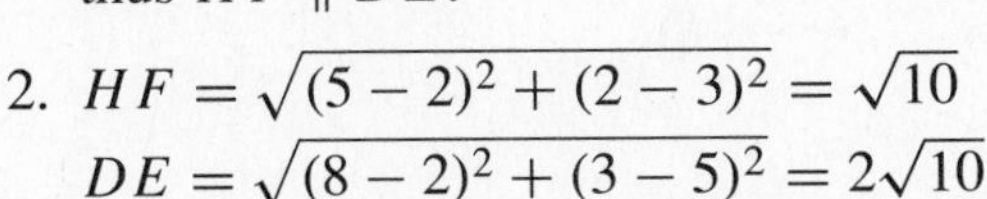

2. $HF = \sqrt{(5-2)^2 + (2-3)^2} = \sqrt{10}$

   $DE = \sqrt{(8-2)^2 + (3-5)^2} = 2\sqrt{10}$

   The length of $\overline{HF}$ is half the length of $\overline{DE}$.

b. Use the figure at the right to complete each statement.

1. If $QR = 2$, then $XY = \boxed{4}$.
2. $\overline{PR} \parallel \boxed{\overline{XZ}}$
3. If $XZ = 8$, then $PR = \boxed{4}$.
4. $PQ = \frac{1}{2}\boxed{YZ}$

c. Given $CB = 5$, $AB = 7$, and $BL = 3$, find the perimeter of $\triangle JKL$.

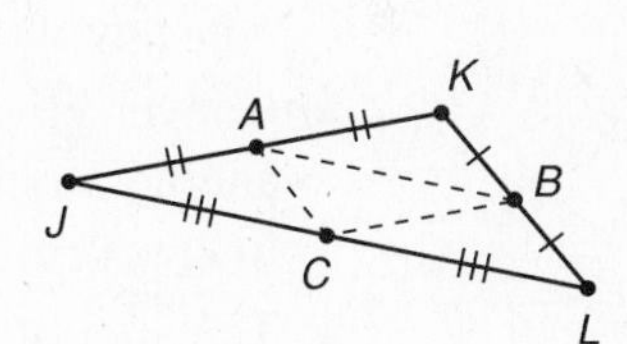

If $CB = 5$, then $JK = 10$. If $BL = 3$, then $KL = 6$.

If $AB = 7$, then $JL = 14$.

The perimeter of $\triangle JKL$ is $10 + 6 + 14 = 30$.

Guidelines:

- A *midsegment* of a triangle is a segment that connects the midpoints of two sides of the triangle.
- The segment connecting the midpoints of two sides of a triangle is parallel to the third side and is half its length.

EXERCISE

The midpoints of the sides of a triangle are $P(2, 1)$, $Q(5, 2)$, and $R(2, -2)$. What are the coordinates of the vertices of the triangle? (Hint: Graph paper may be helpful.)

Reteach
Chapter 5

Name ______________________

What you should learn:

5.4	How to compare the measures of a triangle and use the Triangle Inequality to solve problems

Correlation to Pupil's Textbook:

Chapter Test (p. 263)
Exercises 13–16, 18, 20

Examples *Comparing Measures of a Triangle and Using the Triangle Inequality*

a. Use the figure below to state the longest and shortest sides of the triangle.

Longest side: $\overline{JL}$ Shortest side: $\overline{JK}$

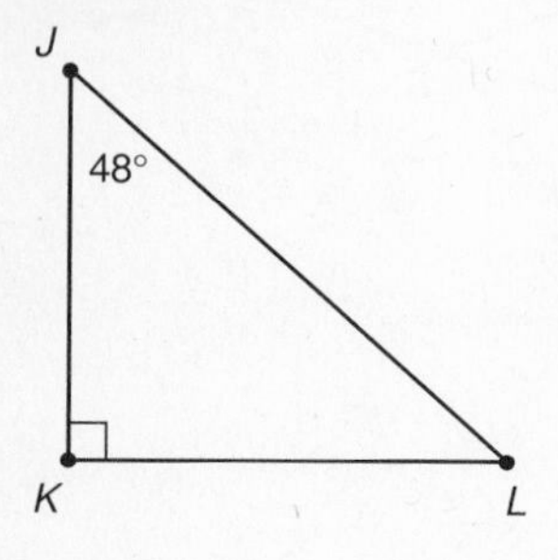

b. Use the figure below to state the largest and smallest angles of the triangle.

Largest angle: $\angle S$ Smallest angle: $\angle T$

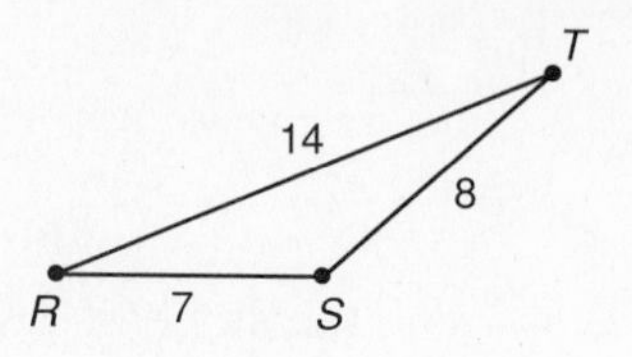

c. Two sides of a triangle measure 10 and 15. Describe the measure of the third side x. If $10 + x > 15$, then $x > 5$. Also $10 + 15 > x$, therefore, $5 < x < 25$.

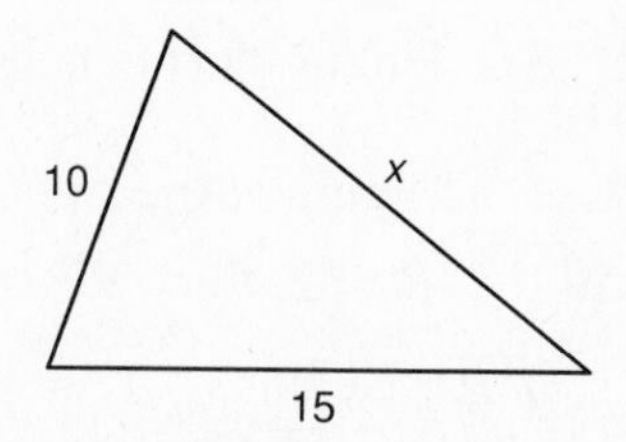

Guidelines:

- If one side of a triangle is longer than another side, then the angle opposite the longer side is larger than the angle opposite the shorter side.
- Conversely, if one angle of a triangle is larger than another angle, then the side opposite the larger angle is longer than the side opposite the smaller angle.
- The sum of the lengths of any two sides of a triangle is greater than the length of the third side.

EXERCISES

1. Order the sides from shortest to longest.

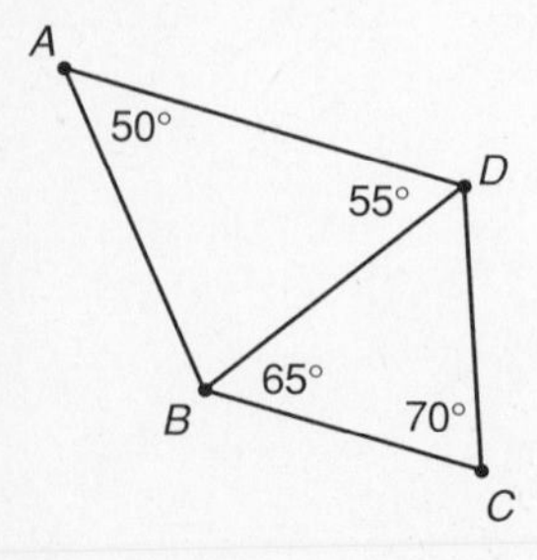

2. Order the angles from smallest to largest.

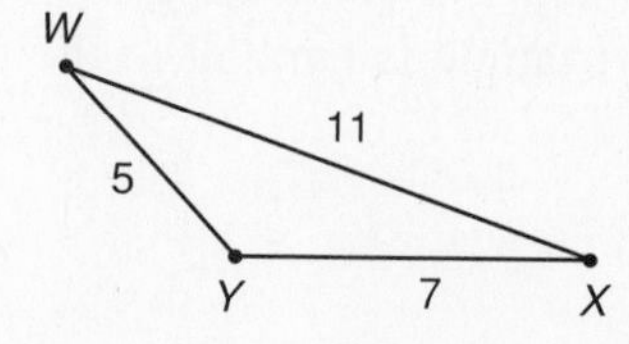

Reteach
Chapter 5

Name ______________________

What you should learn:

5.5	How to use an indirect proof and use the Hinge Theorem

Correlation to Pupil's Textbook:

Chapter Test (p. 263)
Exercises 14, 17, 19

Examples *Using Indirect Proof and Using the Hinge Theorem*

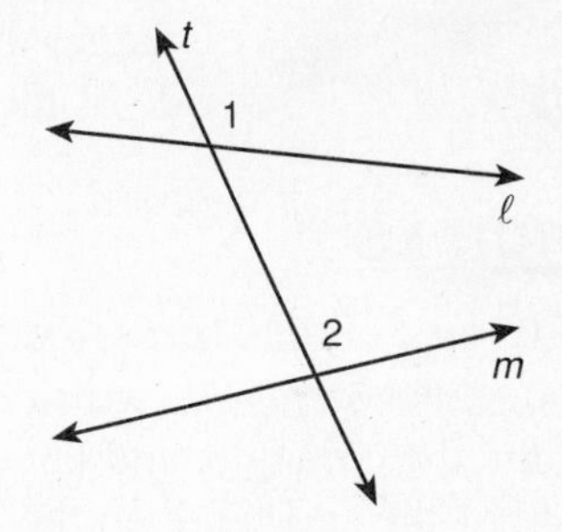

a. Write an indirect proof.

Given: Lines ℓ and m are cut by a transversal t.

$\angle 1 \ncong \angle 2$

Prove: ℓ and m are not parallel.

Suppose that ℓ and m are parallel. Then corresponding angles $\angle 1$ and $\angle 2$ are congruent. This contradicts the given statement that $\angle 1 \ncong \angle 2$. The case that ℓ and m are parallel is not true and the case that ℓ and m are not parallel is true.

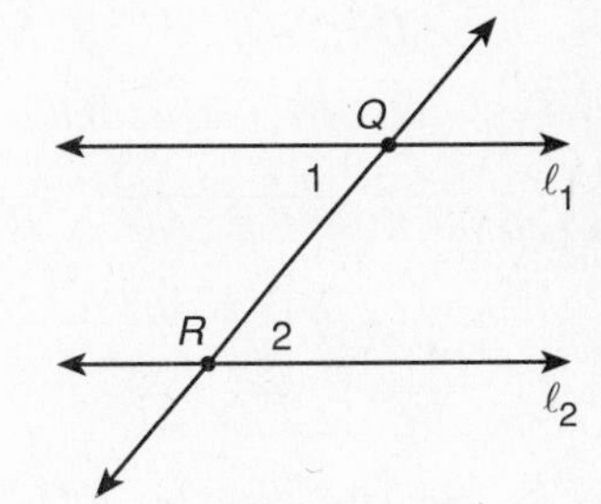

b. Write an indirect proof for the Alternate Interior Angles Converse Theorem—If two lines are cut by a transversal so that alternate interior angles are congruent, then the lines are parallel.

Given: Lines ℓ_1 and ℓ_2 are cut by transversal t.

$\angle 1 \cong \angle 2$

Prove: $\ell_1 \parallel \ell_2$

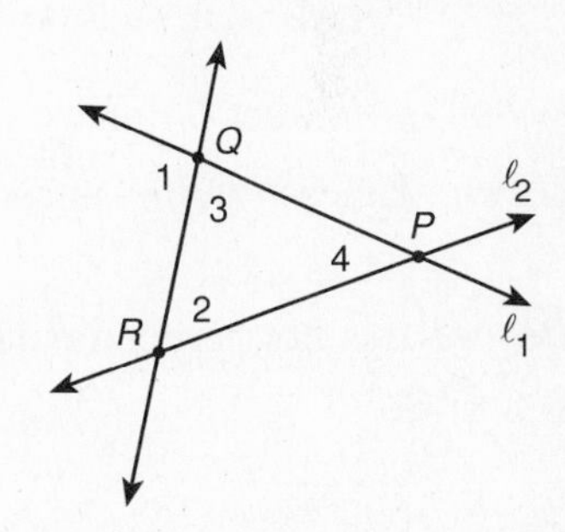

Suppose the lines are not parallel. Then ℓ_1 and ℓ_2 intersect at point P. By the linear pair postulate, $m\angle 1 + m\angle 3 = 180°$. In triangle $\triangle PQR, m\angle 2 + m\angle 3 + m\angle 4 = 180°$. Therefore, $m\angle 1 = m\angle 2 + m\angle 4$. This contradicts the given statement that alternate interior angles $\angle 1 \cong \angle 2$. The case that lines ℓ_1 and ℓ_2 are not parallel is not true and the case that lines ℓ_1 and ℓ_2 are parallel is true.

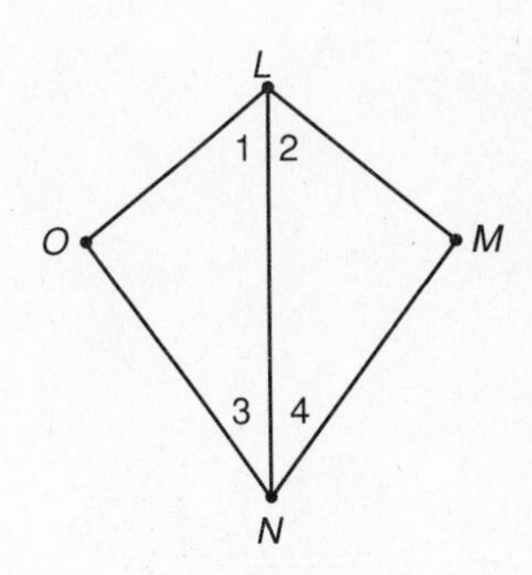

c. Use the Hinge Theorem and its converse to complete the following statements for the given figure.

1. If $\overline{OL} \cong \overline{ML}$ and $m\angle 2 > m\angle 1$, then $\boxed{MN > ON}$.

2. If $\overline{ON} \cong \overline{MN}$ and $OL > ML$, then $\boxed{m\angle 3 > m\angle 4}$.

Reteach
Chapter 5

Name ____________________________

Guidelines:

- With an indirect proof (also known as proof by contradiction), you argue that, of all possible cases, all but one is impossible. Then the remaining case must be true.
- Hinge Theorem—If two sides of one triangle are congruent to two sides of another triangle, and the included angle of the first is larger than the included angle of the second, then the third side of the first is longer than the third side of the second.

EXERCISES

1. Give an example (draw a diagram) that illustrates the Converse of the Hinge Theorem—If two sides of one triangle are congruent to two sides of another triangle, and the third side of the first is longer than the third side of the second, then the included angle of the first is larger than the included angle of the second.

In Exercises 2–4, provide a conclusion from the given information.

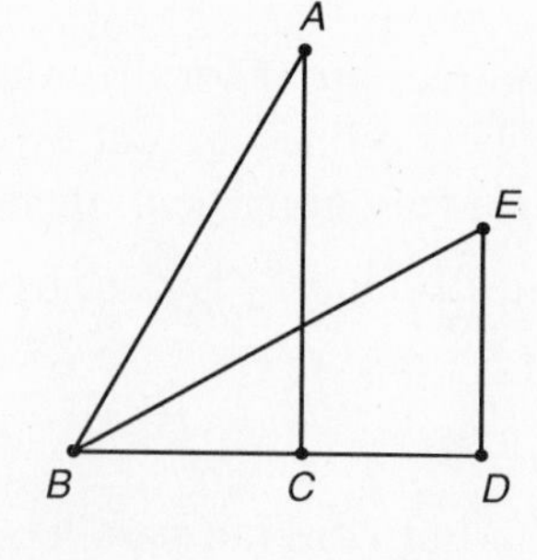

2. *Given:* $\overline{AB} \cong \overline{BE}$, $\overline{BC} \cong \overline{ED}$

$m\angle ABC > m\angle BED$

3. *Given:* $\overline{AB} \cong \overline{BE}$, $\overline{AC} \cong \overline{BD}$

$m\angle EBD > m\angle BAC$

4. *Given:* $\overline{BC} \cong \overline{ED}$, $\overline{AC} \cong \overline{BD}$

$m\angle BCA > m\angle EDB$

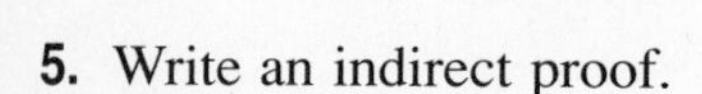

5. Write an indirect proof.

Given: Lines ℓ and m intersect each other.

$\angle 1 \ncong \angle 2$

Prove: ℓ is not perpendicular to m.

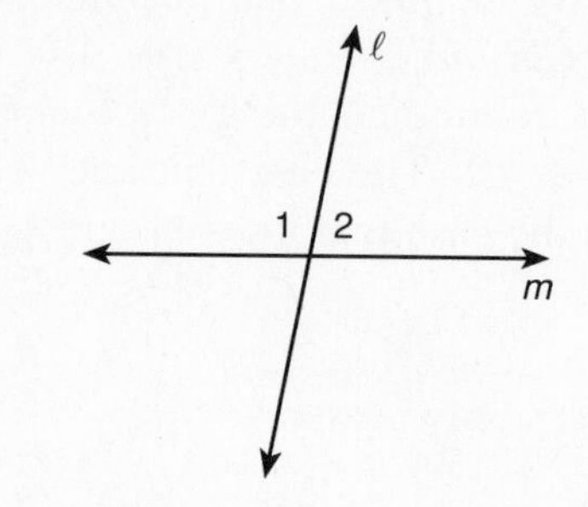

Reteach

Chapter 5

Name ______________________________

What you should learn:

5.6	How to use standard construction techniques and computer constructions

Examples *Using Standard Construction Techniques and Computer Constructions*

a. Use a straightedge and compass to locate the centroid of the given triangle $\triangle XYZ$.

Bisect each side of the triangle to locate midpoints M_1, M_2, and M_3. Construct each median by connecting each vertex of triangle $\triangle XYZ$ with the midpoint of the opposite side. The point of intersection of the medians is the centroid, point P.

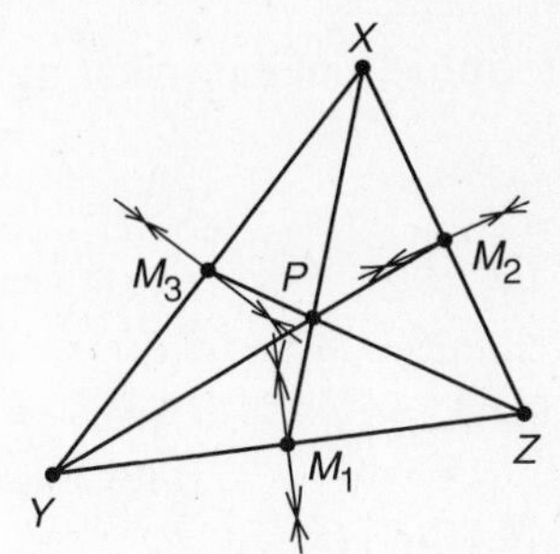

b. Use a computer drawing program to draw a triangle that has an angle bisector that is also a perpendicular bisector.

You can use a computer drawing program to construct an isosceles triangle $\triangle ABC$. The angle bisector, $\overline{AD}$, of $\angle A$ is also a perpendicular bisector of $\overline{BC}$.

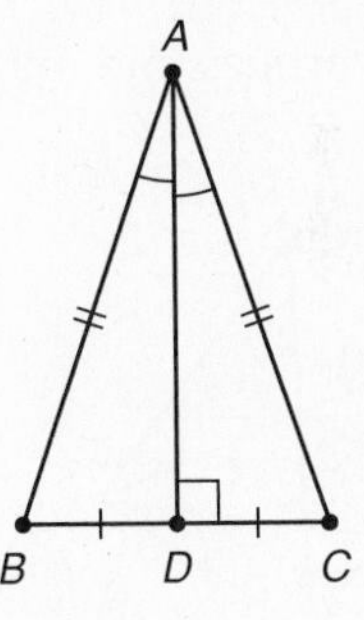

Guidelines:

- A computer drawing program can be used to make constructions that are too difficult to make using standard construction techniques.

EXERCISES

1. Use a straightedge and compass to construct the circumcenter, C, of right triangle $\triangle PQR$. Construct the circle through P whose center is C. What is the name of this circle?

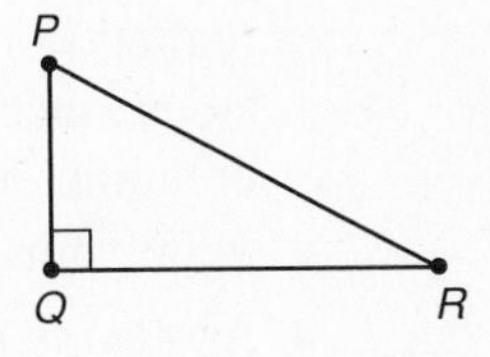

2. Use a computer drawing program to draw a triangle that has an altitude that is also a perpendicular bisector.

Reteach

Chapter 6

Name ______________________________

What you should learn:

6.1	How to identify polygons and use polygons to solve real-life problems

Correlation to Pupil's Textbook:

Mid-Chapter Self-Test (p. 286)
Exercises 1–4, 12

Chapter Test (p. 317)
Exercises 1–6, 9

Examples *Identifying Polygons and Using Polygons in Real Life*

a. Determine whether each figure is a polygon. If it is a polygon, is it convex?

Not a polygon

Not a polygon

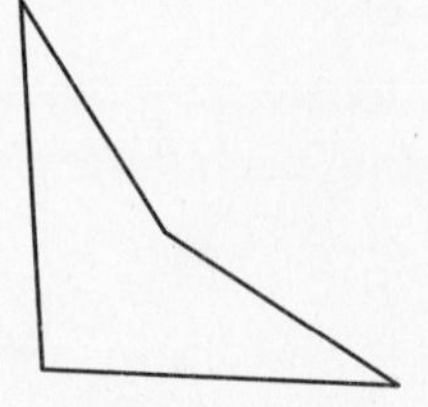
Nonconvex polygon

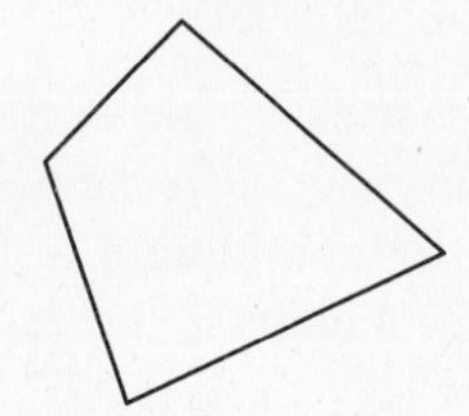
Convex polygon

b. Name the vertices and diagonals of pentagon $ABCDE$.

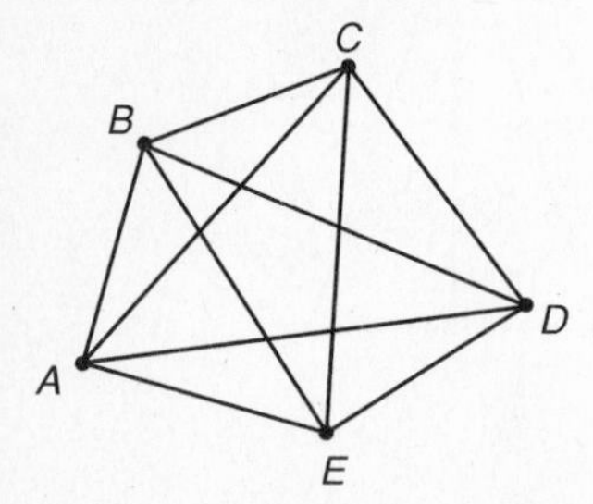

Vertices: A, B, C, D, E
Diagonals: $\overline{AC}, \overline{AD}, \overline{BE}, \overline{BD}, \overline{CE}$

c. A design for a logo is created by joining together five congruent isosceles triangles as shown. Classify the polygon as equilateral, equiangular, or regular.

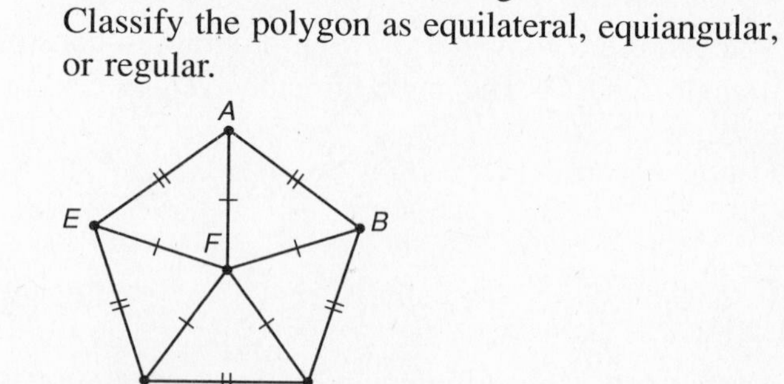

The polygon is a regular pentagon.

Guidelines:

- A *polygon* is a plane figure formed by three or more segments (called sides) such that each side intersects exactly two other sides, once at each endpoint, and such that no two sides with a common endpoint are collinear.
- A polygon is named by listing its vertices consecutively.
- A *regular polygon* is both equilateral and equiangular.

EXERCISES

In Exercises 1–4, classify the polygons as equilateral, equiangular, or regular. Name the vertices and diagonals of each polygon.

1.

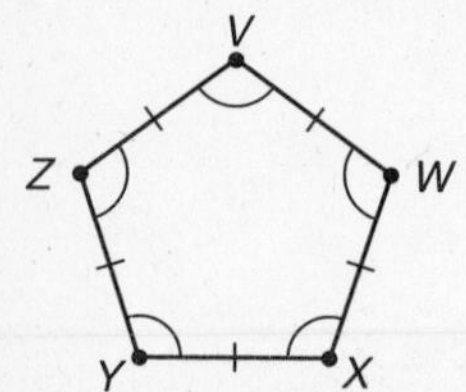

2.

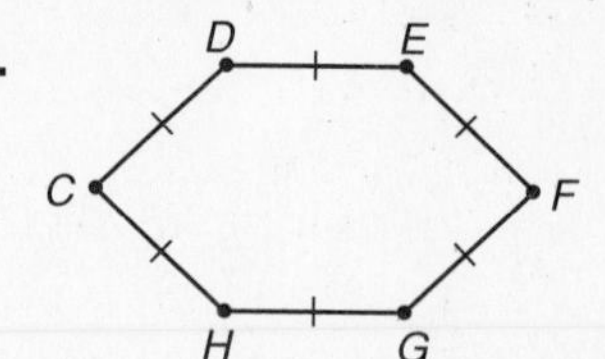

3.

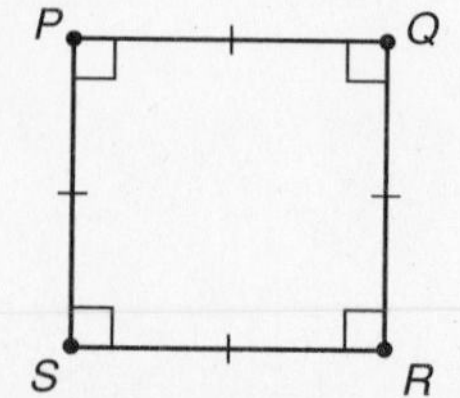

4.

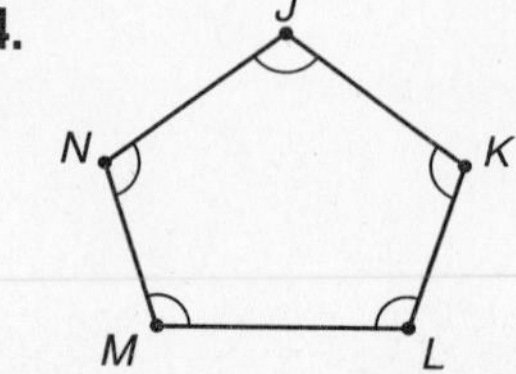

Reteach

Chapter 6

Name ______________________________

What you should learn:

6.2 How to find angles measures of polygons and use angle measures to solve real-life problems

Correlation to Pupil's Textbook:

Mid-Chapter Self-Test (p. 286) Exercises 5–10, 13

Chapter Test (p. 317) Exercises 7, 8

Examples *Finding Angles Measures of Polygons and Using Angles Measures in Real Life*

a. Find the measure of each interior angle and each exterior angle of a regular dodecagon.

Since a dodecagon has 12 sides, $n = 12$. Using the Polygon Interior Angles Theorem for $n = 12$, the sum of the measures of the interior angles is $(12 - 2)(180°) = 1800°$. Since the polygon is regular, the measure of each interior angle is $\frac{1800}{12} = 150°$. The measure of each exterior angle of regular n-gon is $\frac{1}{n}(360°)$. When $n = 12$, the measure is $\frac{1}{12}(360°) = 30°$. Note that the interior and exterior angles are linear pairs.

b. The measure of each interior angle of a regular polygon is 168°. The polygon has how many sides?

Set 168° equal to the formula for the measure of each interior angle and solve for n.

$$\frac{1}{n}(n - 2)(180°) = 168°$$

$$180n - 360 = 168n$$

$$12n = 360$$

$$n = 30$$

The polygon has 30 sides.

c. You have designed a flower garden, as shown by quadrilateral $ABCD$ at the right. The measure of angle $\angle A$ is twice the measure of angle $\angle D$ which is twice the measure of $\angle C$. If the measure of $\angle B$ is 115°, find the measures of angles A, C, and D.

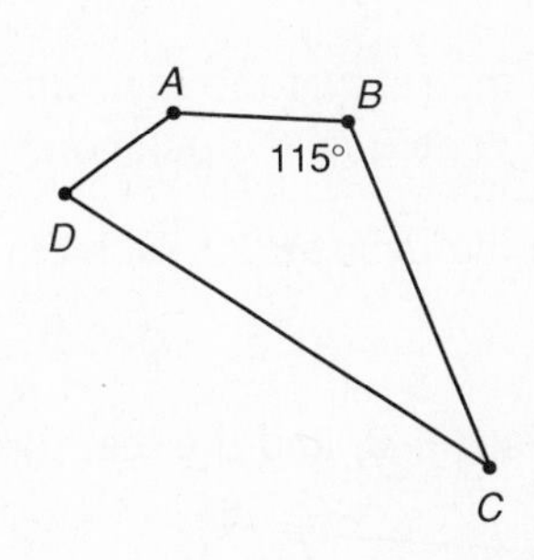

Let $x = m\angle C$, then $2x = m\angle D$ and $4x = m\angle A$. Using the Polygon Interior Angles Theorem for $n = 4$, the sum of the measures of the interior angles is $(4 - 2)180° = 360°$. When $x + 2x + 4x + 115° = 360°$, $x = 35°$. Therefore, $m\angle C = 35°$, $m\angle D = 70°$, $m\angle A = 140°$.

Reteach
Chapter 6

Name ____________________

d. Using the figure at the right, show that the sum of the measures of the exterior angles, one from each vertex, of a convex polygon is 360°.

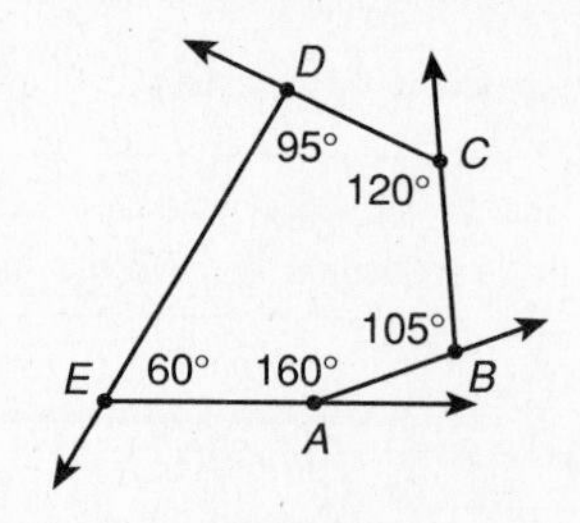

By the Linear Pair Postulate, each exterior angle is supplementary to its corresponding interior angle. The measures of the exterior angles are 20°, 75°, 60°, 85° and 120°. The sum of the measures of the exterior angles is 360°.

Guidelines:

- The sum of the measures of the interior angles of a convex n-gon is $(n - 2)(180°)$.
- The measure of each interior angle of a regular n-gon is $\frac{1}{n}(n - 2)(180°)$.
- The sum of the measures of the exterior angles, one from each vertex, of a convex polygon is 360°.
- The measure of each exterior angle of a regular n-gon is $\frac{1}{n}(360°)$.

EXERCISES

1. Two diagonals of a regular pentagon form three triangles, as shown at the right. Find the measure of all numbered angles.

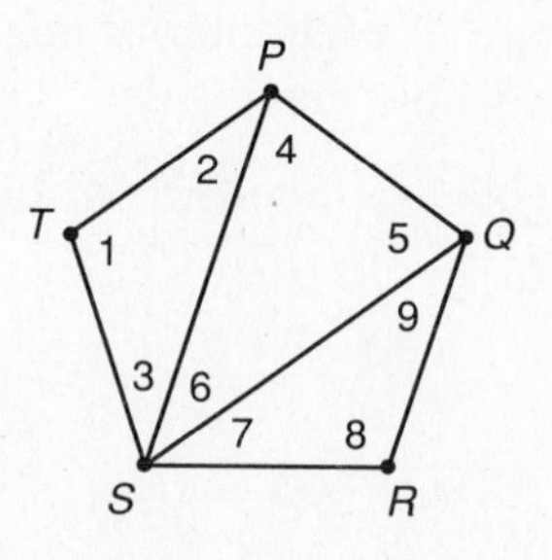

In Exercises 2–5, consider a regular octagon as shown at the right.

2. Find the sum of the measures of the interior angles of the octagon.

3. Find the measure of each interior angle of the regular octagon.

4. Find the sum of the measures of the exterior angles, one from each vertex, of the octagon.

5. Find the measure of each exterior angle of the regular octagon.

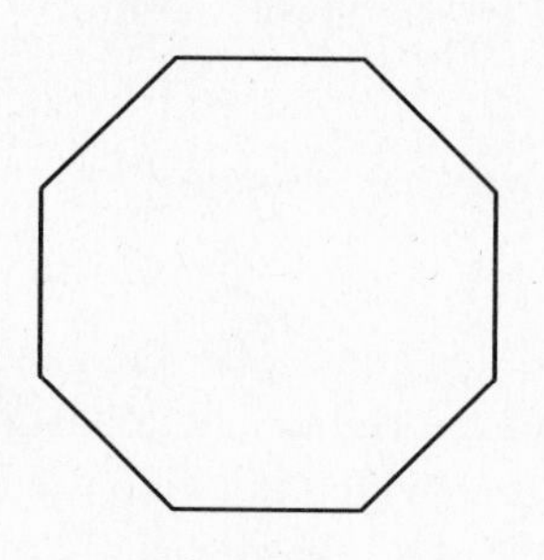

In Exercises 6–8, find the measure of $\angle A$.

6.

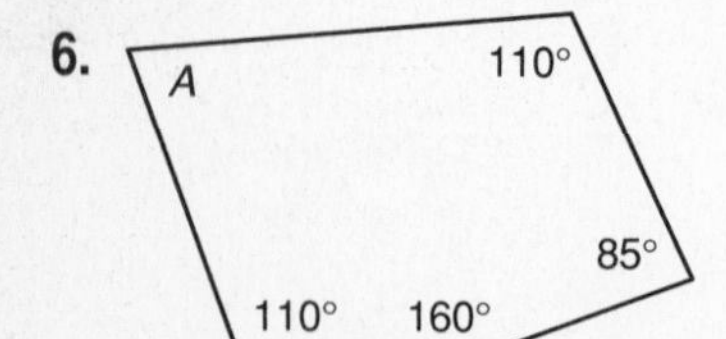

7.

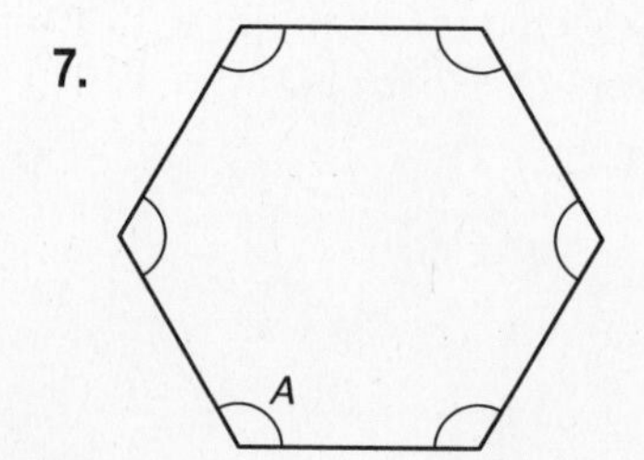

8.

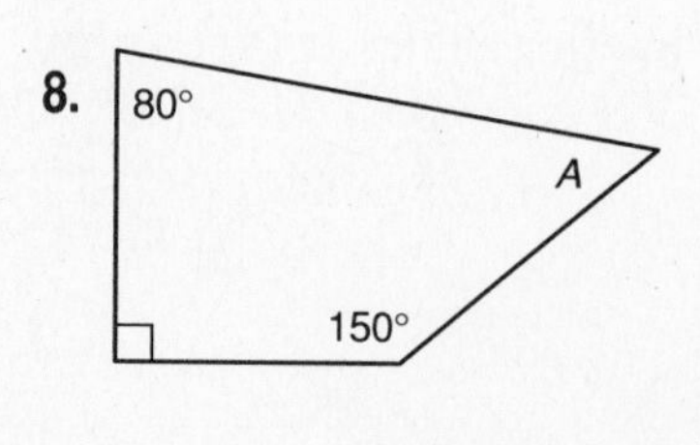

Reteach
Chapter 6

Name ______________________

What you should learn:

6.3	How to use properties of parallelograms to solve problems in geometry and in real life

Correlation to Pupil's Textbook:

Mid-Chapter Self-Test (p. 286) Exercises 11, 14–20

Chapter Test (p. 317) Exercise 13

Examples *Using Properties of Parallelograms and Using Parallelograms in Real Life*

a. Use the diagram of $\square RSTU$ at the right to list (1) congruent angles, (2) congruent sides, and (3) supplementary angles.

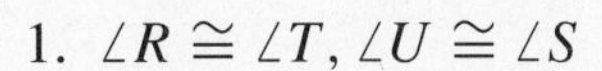

1. $\angle R \cong \angle T$, $\angle U \cong \angle S$
2. $\overline{RS} \cong \overline{TU}$, $\overline{RU} \cong \overline{TS}$
3. $m\angle R + m\angle S = 180^\circ$
   $m\angle S + m\angle T = 180^\circ$
   $m\angle T + m\angle U = 180^\circ$
   $m\angle U + m\angle R = 180^\circ$

b. Use the diagram of $\square DEFG$ to match each of the following.

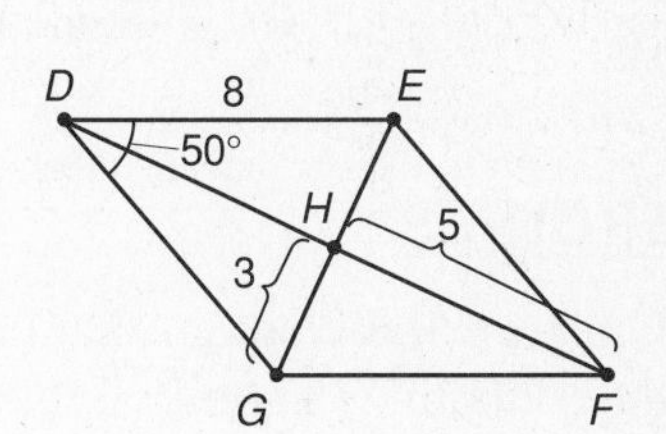

1. DF a. 8
2. GF b. 10
3. HE c. 3
4. $m\angle EFG$ d. 130°
5. $m\angle DEF$ e. 50°

Answers: 1. b 2. a 3. c 4. e 5. d

c. Use the diagram of $\square JKLM$ at the right and the Distance Formula to show that opposite sides are congruent.

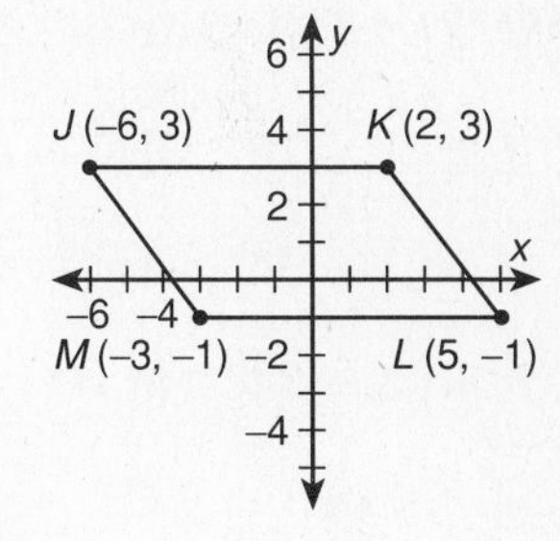

Using the Distance Formula,

$JK = \sqrt{(-6-2)^2 + (3-3)^2} = \sqrt{(-8)^2 + 0^2} = \sqrt{64} = 8$

and

$ML = \sqrt{(-3-5)^2 + (-1-(-1))^2} = \sqrt{(-8)^2 + 0^2} = \sqrt{64} = 8.$

Therefore $\overline{JK} \cong \overline{ML}$.

Using the Distance Formula again,

$JM = \sqrt{(-6-(-3))^2 + (3-(-1))^2} = \sqrt{(-3)^2 + 4^2} = \sqrt{9+16} = \sqrt{25} = 5$

and

$KL = \sqrt{(2-5)^2 + (3-(-1))^2} = \sqrt{(-3)^2 + 4^2} = \sqrt{9+16} = \sqrt{25} = 5$

Therefore $\overline{JM} \cong \overline{KL}$.

Name ______________________________

d. Chevron symbols, as shown at the right, are used to direct traffic flow on city streets. If $ABCD$ and $CDEF$ are parallelograms, what can you conclude about $\overline{AB}$ and $\overline{EF}$?

Since quadrilateral $ABCD$ is a parallelogram, $\overline{AB} \cong \overline{CD}$ and $\overline{AB} \parallel \overline{CD}$. Also quadrilateral $CDEF$ is a parallelogram. Then $\overline{CD} \cong \overline{EF}$ and $\overline{CD} \parallel \overline{EF}$. By the Transitive Property of Congruence and Transitivity of Parallel Lines, $\overline{AB} \cong \overline{EF}$ and $\overline{AB} \parallel \overline{EF}$.

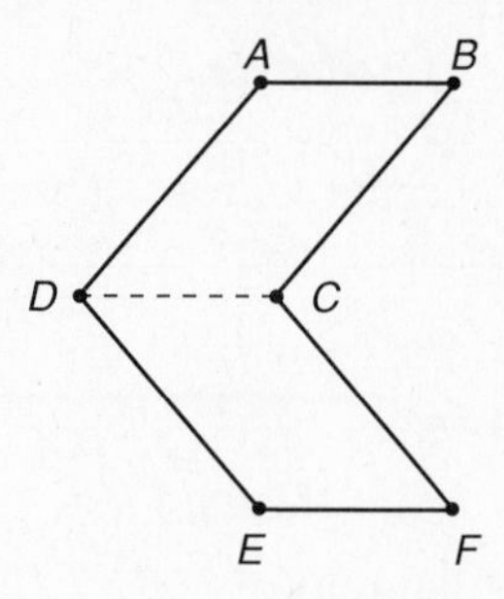

Guidelines: If a quadrilateral is a parallelogram, then

- opposite sides are congruent.
- opposite angles are congruent.
- opposite sides are parallel.
- consecutive angles are supplementary.
- diagonals bisect each other.

EXERCISES

1. Draw a parallelogram that has a right angle. What can you conclude about the remaining angles? Explain your reasoning.

In Exercises 2–8, use the diagram of $\square ABCD$ at the right.

2. If $AC = 12$, what is EC?

3. If $AD = 14$, what is CB?

4. If $m\angle 1 = 10°$, what is $m\angle 2$?

5. If $m\angle 3 = 34°$, what is $m\angle 4$?

6. If $m\angle 3 = 34°$ and $m\angle 1 = 10°$, what is $m\angle ABC$?

7. If $m\angle DAB = 132°$, what is $m\angle BCD$?

8. If $m\angle DAB = 132°$, what is $m\angle ABC$?

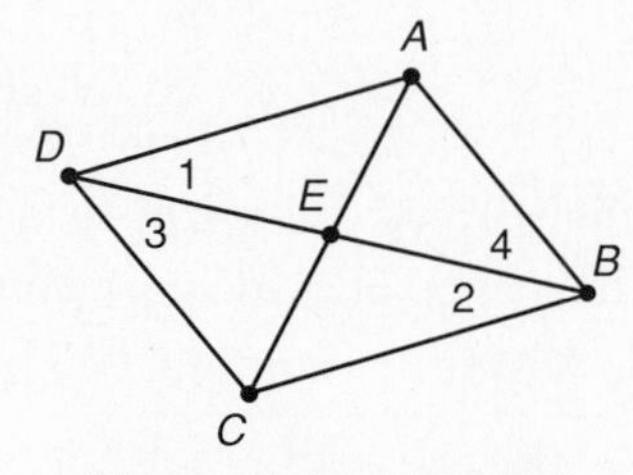

(Figure not drawn to scale)

9. Use the figure at the right. If $ABCG$ and $CDEF$ are parallelograms, what can you conclude about $\angle GAB$ and $\angle DEF$? Explain your reasoning.

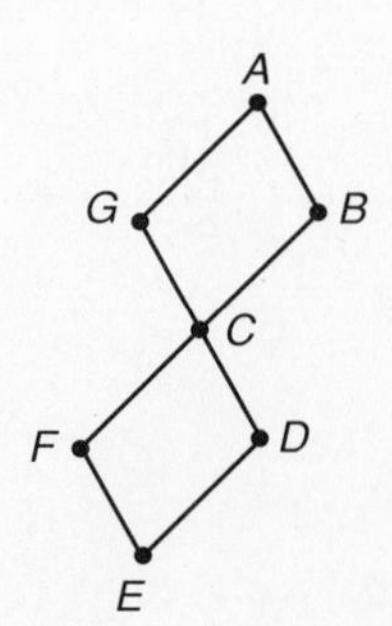

Reteach
Chapter 6

Name ______________________________

What you should learn:

6.4	How to prove that quadrilaterals are parallelograms and use coordinate geometry

Correlation to Pupil's Textbook:

Chapter Test (p. 317)
Exercise 13

Examples *Proving Parallelograms and Using Coordinate Geometry*

a. Decide whether you are given enough information to determine that $PQRS$, shown below, is a parallelogram. Explain.

1. $\overline{PQ} \cong \overline{SR}$, $\overline{PS} \parallel \overline{QR}$ — 1. No—need 2 pairs of opposite congruent or 2 pairs of opposite parallel sides.
2. $\overline{PQ} \parallel \overline{SR}$, $\overline{PQ} \cong \overline{SR}$ — 2. Yes—because one pair of opposite sides is both congruent and parallel.
3. $m\angle SPQ + m\angle PQR = 180^\circ$, $m\angle SPQ + m\angle RSP = 180^\circ$ — 3. Yes—because an angle is supplementary to both of its consecutive angles.
4. $\overline{PQ} \cong \overline{RS}$, $\overline{PS} \cong \overline{RQ}$ — 4. Yes—because both pairs of opposite sides are congruent.
5. $\overline{PT} \cong \overline{RT}$, $\overline{ST} \cong \overline{QT}$ — 5. Yes—because the diagonals bisect each other.
6. $\angle PTS \cong \angle QTR$, $\angle PTQ \cong \angle RTS$ — 6. No—need pairs of opposite angles to be congruent, not vertical angles.

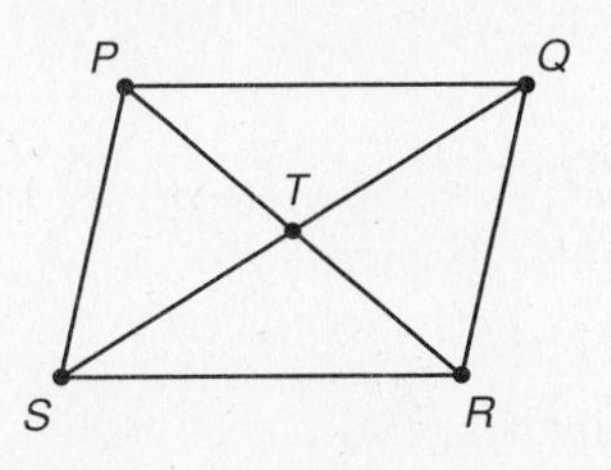

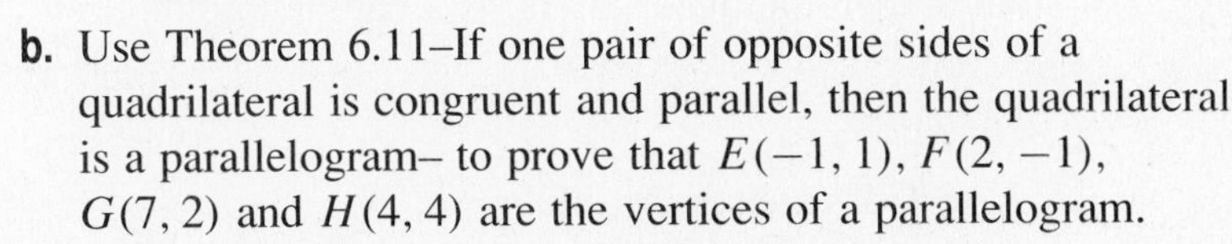

b. Use Theorem 6.11–If one pair of opposite sides of a quadrilateral is congruent and parallel, then the quadrilateral is a parallelogram– to prove that $E(-1, 1)$, $F(2, -1)$, $G(7, 2)$ and $H(4, 4)$ are the vertices of a parallelogram.

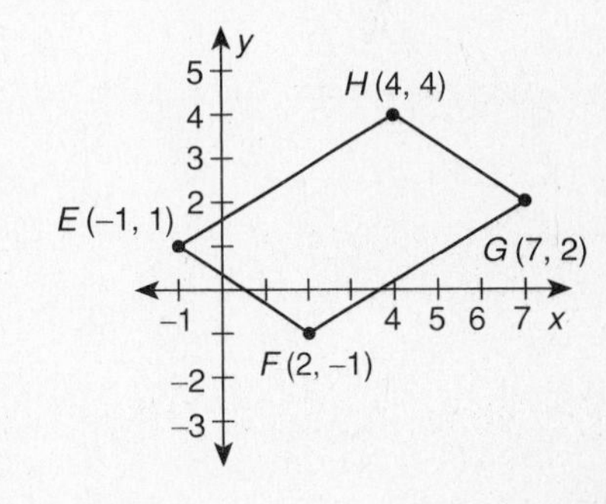

You can show that $\overline{EH} \cong \overline{FG}$ and $\overline{EH} \parallel \overline{FG}$.

$$EH = \sqrt{(4-(-1))^2 + (4-1)^2} = \sqrt{34}$$

$$\text{Slope of } \overline{EH} = \frac{4-1}{4+1} = \frac{3}{5}$$

$$FG = \sqrt{(7-2)^2 + (2+1)^2} = \sqrt{34}$$

$$\text{Slope of } \overline{FG} = \frac{2+1}{7-2} = \frac{3}{5}$$

Since $\overline{EH} \cong \overline{FG}$ and $\overline{EH} \parallel \overline{FG}$, quadrilateral $EFGH$ is a parallelogram.

Reteach
Chapter 6

Name ______________________________

Guidelines: You can prove that a quadrilateral is a parallelogram by proving that

- both pairs of opposite sides are congruent.
- both pairs of opposite angles are congruent.
- diagonals bisect each other.
- an angle is supplementary to both of its consecutive angles.
- one pair of opposite sides is both parallel and congruent.

EXERCISES

1. Rework Example b on page 53 using Theorem 6.7–If both pairs of opposite sides of a quadrilateral are congruent, then the quadrilateral is a parallelogram.

2. Match the correct reason for each step of the proof.
Given: $\overline{EH} \cong \overline{GF}$, $\overline{EH} \parallel \overline{GF}$
Prove: $EFGH$ is a parallelogram.

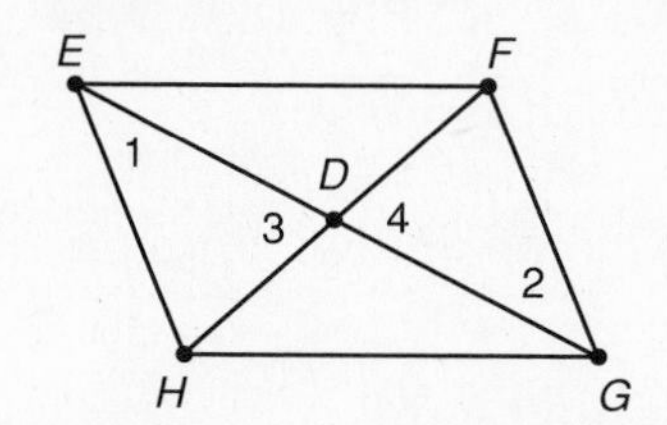

Proof:

Statements	*Reasons*
1. $\overline{EH} \parallel \overline{GF}$	**1.** ?
2. $\angle 1 \cong \angle 2$	**2.** ?
3. $\angle 3 \cong \angle 4$	**3.** ?
4. $\overline{EH} \cong \overline{GF}$	**4.** ?
5. $\triangle EDH \cong \triangle GDF$	**5.** ?
6. $\overline{ED} \cong \overline{GD}$	**6.** ?
7. $\overline{DH} \cong \overline{DF}$	**7.** ?
8. $EFGH$ is a ▱.	**8.** ?

a. CPCTC
b. Vertical Angles Theorem
c. Alternate Interior Angles Theorem
d. CPCTC
e. If the diagonals of a quadrilateral bisect each other, then the quadrilateral is a parallelogram.
f. Given
g. Given
h. AAS Congruence Postulate

Reteach
Chapter 6

Name ____________________

What you should learn:

6.5 How to identify and use special parallelograms

Correlation to Pupil's Textbook:

Chapter Test (p. 317)
Exercises 14–16

Examples *Identifying and Using Special Parallelograms*

a. Identify each special parallelogram and explain your reasoning. Describe the diagonals of each special parallelogram.

1. 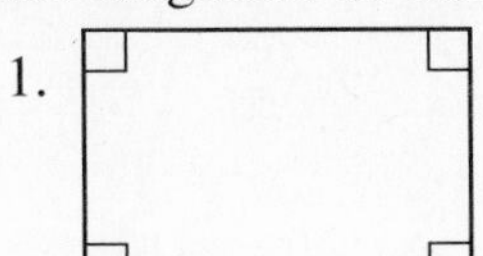

Rectangle: It has four right angles. Diagonals are congruent.

2.

Rhombus: It has congruent sides. Diagonals are perpendicular. Each diagonal bisects a pair of opposite angles.

3.

Square: It has four congruent sides and four right angles. Diagonals are congruent and perpendicular.

b. Write a two-column proof, using the given figure.

Given: $\overline{AB} \cong \overline{BC} \cong \overline{CD} \cong \overline{DA}$

Prove: $\angle 1$ is a right angle.

Proof:

Statements	*Reasons*
1. 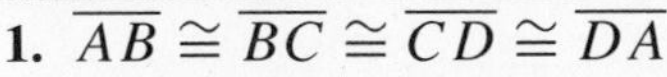 $\overline{AB} \cong \overline{BC} \cong \overline{CD} \cong \overline{DA}$	**1.** Given
2. Quadrilateral $ABCD$ is a rhombus.	**2.** A quadrilateral is a rhombus if and only if it has four congruent sides.
3. $\overline{BD} \perp \overline{AC}$	**3.** A parallelogram is a rhombus if and only if its diagonals are perpendicular.
4. $\angle 1$ is a right angle.	**4.** If two lines are perpendicular, they meet to form a right angle.

Guidelines: By definition,

- a *rhombus* is a parallelogram with four congruent sides.
- a *rectangle* is a parallelogram with four right angles.
- a *square* is a parallelogram that is both a rhombus and a rectangle.

EXERCISE

You are building a wooden trellis for a flower garden, as shown at the right. If each of the three sections is a congruent rhombus and one edge is 9 inches long, how much wood is needed to build the trellis?

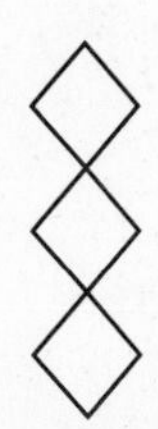

Reteach

Chapter 6

Name ____________________

What you should learn:

6.6	How to identify trapezoids and use properties of trapezoids

Correlation to Pupil's Textbook:

Chapter Test (p. 317)
Exercises 17–19

Examples *Identifying Trapezoids and Using Properties of Trapezoids*

a. Using the figure at the right, identify the bases, consecutive sides, diagonals, base angles, legs, and opposite angles of the trapezoid.

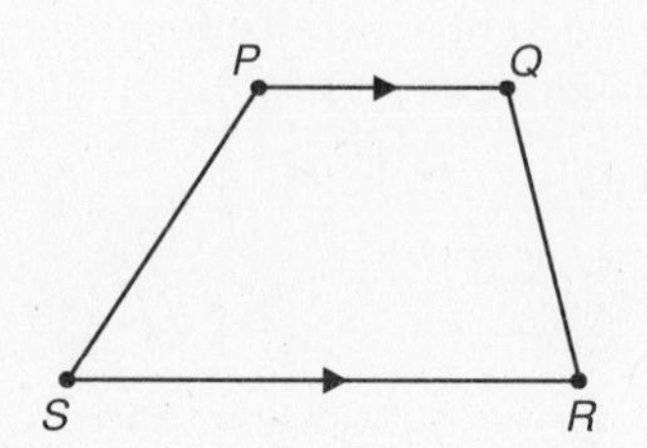

	Answers:
1. Bases	$\overline{PQ}$ and $\overline{SR}$
2. Consecutive sides	$\overline{PQ}$ and $\overline{QR}$, $\overline{QR}$ and $\overline{RS}$, $\overline{RS}$ and $\overline{SP}$, $\overline{SP}$ and $\overline{PQ}$
3. Diagonals	$\overline{SQ}$ and $\overline{PR}$
4. Base angles	$\angle P$ and $\angle Q$, $\angle S$ and $\angle R$
5. Legs	$\overline{PS}$ and $\overline{QR}$
6. Opposite angles	$\angle P$ and $\angle R$, $\angle Q$ and $\angle S$

b. If trapezoid $PQRS$ in Example a is an isosceles trapezoid, which angles are congruent? How are diagonals $\overline{SQ}$ and $\overline{PR}$ related?

If a trapezoid is isosceles, then each pair of base angles is congruent. That is, $\angle S \cong \angle R$ and $\angle P \cong \angle Q$. If a trapezoid is isosceles, then its diagonals are congruent. In this case, $\overline{SQ} \cong \overline{PR}$.

c. Find the length of the midsegment MN of the trapezoid $ABCD$.

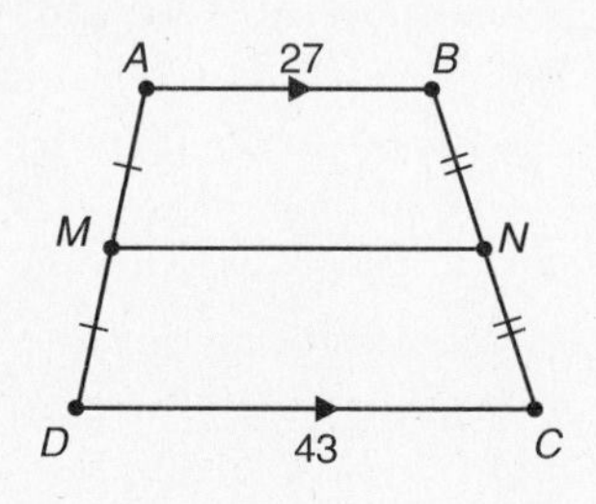

By the Midsegment Theorem for Trapezoids,
$MN = \frac{1}{2}(27 + 43) = 35$.

d. A trapezoid is shown in the figure at the right. Do you have enough information to conclude that the trapezoid is isosceles?

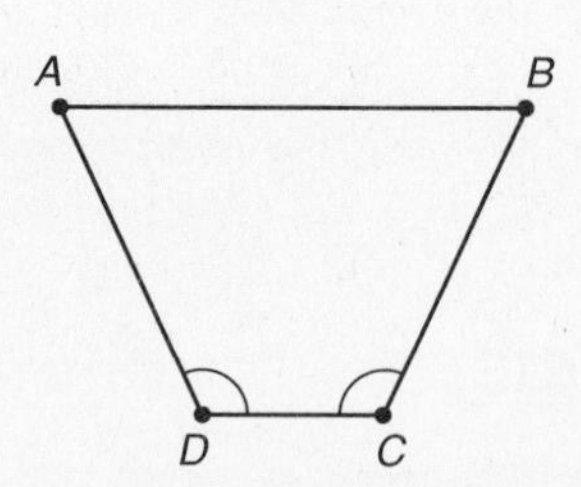

Yes, you can conclude that the trapezoid is isosceles. By Theorem 6.19, if a trapezoid has one pair of congruent base angles, then it is an isosceles trapezoid.

Reteach
Chapter 6

Name ______________________________

Guidelines: By definition,

- A *trapezoid* is a quadrilateral with exactly one pair of parallel opposite sides.
- The *bases of a trapezoid* are the parallel sides.
- The *legs of a trapezoid* are the nonparallel sides.
- If the legs of a trapezoid are congruent, the trapezoid is isosceles.
- A trapezoid has two pairs of base angles.
- If a trapezoid is isosceles, then each pair of base angles is congruent.
- If a trapezoid is isosceles, then its diagonals are congruent.
- The midsegment of a trapezoid is parallel to each base and its length is half the sum of the lengths of the bases.

EXERCISES

1. Using the figure in Example c on page 56, sketch and label the midsegments of trapezoids $ABNM$ and $MNCD$. Find the length of each midsegment.

2. Find $m\angle Z$ and $m\angle Y$ for trapezoid $WXYZ$ shown below.

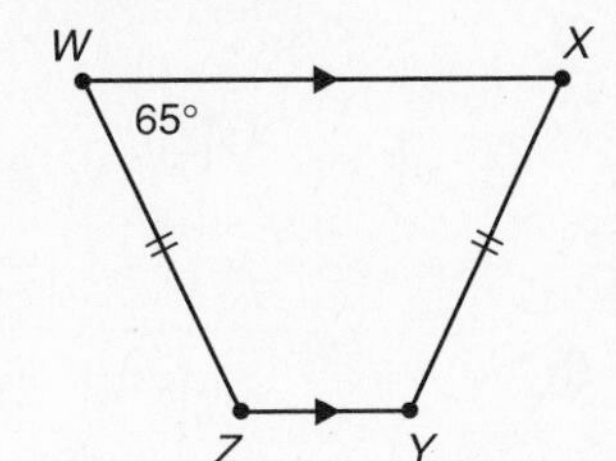

3. Find ON and the length of the midsegment of trapezoid $LMNO$ shown below.

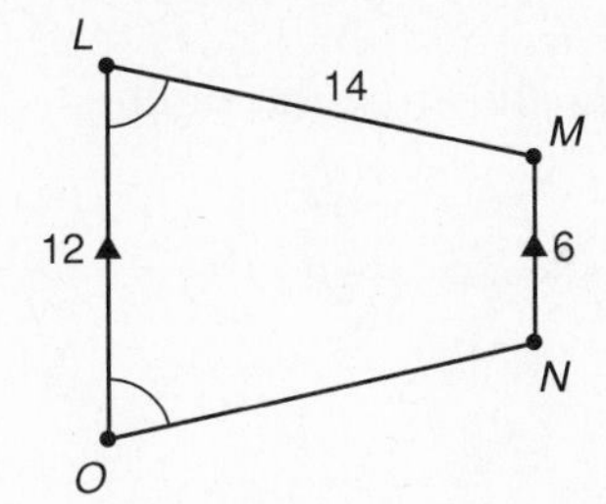

4. Match the correct reason for each step of the proof.
If a trapezoid is isosceles, then its diagonals are congruent.

Given: $\overline{PS} \cong \overline{QR}$, $\overline{PQ} \parallel \overline{SR}$

Prove: $\overline{SQ} \cong \overline{RP}$

Proof:

Statements	*Reasons*
1. $\overline{PQ} \parallel \overline{SR}$	**1.** ?
2. $\overline{PS} \cong \overline{QR}$	**2.** ?
3. $\angle SPQ \cong \angle RQP$	**3.** ?
4. $\overline{PQ} \cong \overline{PQ}$	**4.** ?
5. $\triangle SPQ \cong \triangle RQP$	**5.** ?
6. $\overline{SQ} \cong \overline{RP}$	**6.** ?

a. Base angles of an isosceles triangle are congruent.
b. Given
c. CPCTC
d. Given
e. Reflexive Property of Congruence
f. SAS Congruence Postulate

Reteach
Chapter 6

Name ______________________________

What you should learn:

6.7	How to prove that quadrilaterals are congruent and identify kites

Correlation to Pupil's Textbook:

Chapter Test (p. 317)
Exercises 10–12, 20

Examples *Proving Quadrilaterals Congruent and Identifying Kites*

a. $MNOP$ and $WXYZ$ are rhombuses. Use the given measures to write a paragraph proof that the rhombuses are congruent.

Given: $\overline{MP} \cong \overline{WZ}$, $\angle M \cong \angle W$

Prove: $MNOP \cong WXYZ$

Proof:

- You know that a rhombus has four congruent sides. Since $\overline{MP} \cong \overline{WZ}$, all sides of both rhombuses are congruent.
- Since a rhombus is a parallelogram, $m\angle M + m\angle N = 180°$ and $m\angle W + m\angle X = 180°$. You know that $\angle M \cong \angle W$; therefore, $\angle N \cong \angle X$.
- You can use the SASAS Congruence Theorem to prove that the rhombuses $MNOP$ and $WXYZ$ are congruent.

Remember that just knowing that the sides of both rhombuses are congruent is not sufficient to prove the rhombuses are congruent. SSSS is not a valid method of proving quadrilaterals congruent.

b. Quadrilateral $ABCD$ shown at the right is a kite. Match the following:

1. $\overline{AC}$ [?] $\overline{BD}$ a. $\cong$
2. $\overline{AD}$ [?] $\overline{AB}$ b. $\ncong$
3. $\overline{AD}$ [?] $\overline{CB}$ c. $\perp$

Answers: c, a, b

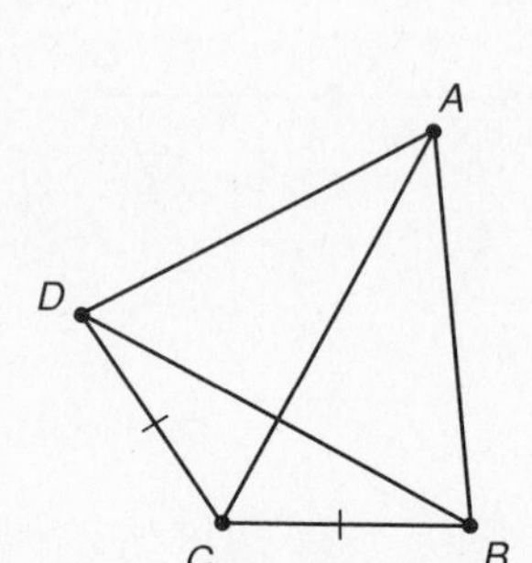

Guidelines: By definition,

- Two quadrilaterals are congruent if their corresponding angles and corresponding sides are congruent.
- A quadrilateral is a kite if it has two pairs of consecutive congruent sides, but opposite sides are not congruent.

EXERCISES

1. How would you change the proof in Exercise a above, if you wished to use the ASASA Congruence Theorem?
2. Since a kite is not a rhombus, it cannot have four congruent sides. What other property of a rhombus can a kite *not* have?

Reteach
Chapter 7

Name ______________________

What you should learn:

7.1	How to identify the three basic rigid transformations and use transformations in real life

Correlation to Pupil's Textbook:

Mid-Chapter Self-Test (p. 344)
Exercises 1, 2, 5, 7–9

Chapter Test (p. 369)
Exercises 1–4

Examples *Identifying Transformations and Using Transformations in Real Life*

a. Identify the image and preimage for the transformation shown at the right. Is the transformation rigid?

The preimage is the figure on the left. The rotation maps the preimage to the image on the right. The transformation is rigid because the image is congruent to its preimage.

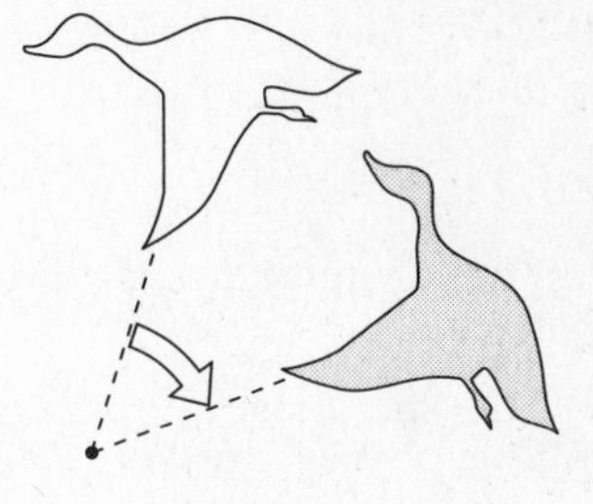

b. An isometry is a transformation in the plane that preserves length. Identify each transformation and each isometry. (Preimages are unshaded and images are shaded.)

Nonrigid transformation
Not an isometry

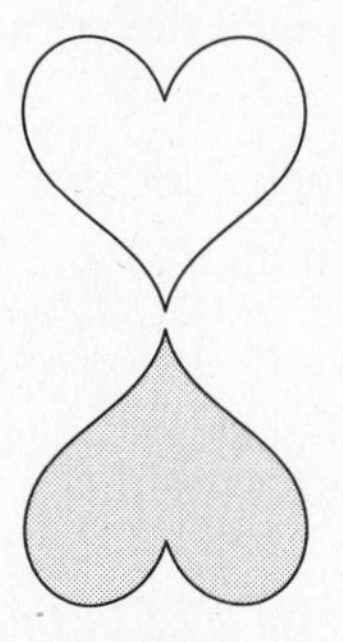

Reflection
Isometry

Rotation
Isometry

c. Name the transformation shown at the right. (Dashed lines denote preimage; solid lines denote image.)

To obtain rectangle $A'B'C'D'$, each point of rectangle $ABCD$ was moved 6 units to the left and 3 units down. This transformation is a translation. The length of each segment is preserved because it is an isometry. Angle measure and parallel lines are also preserved.

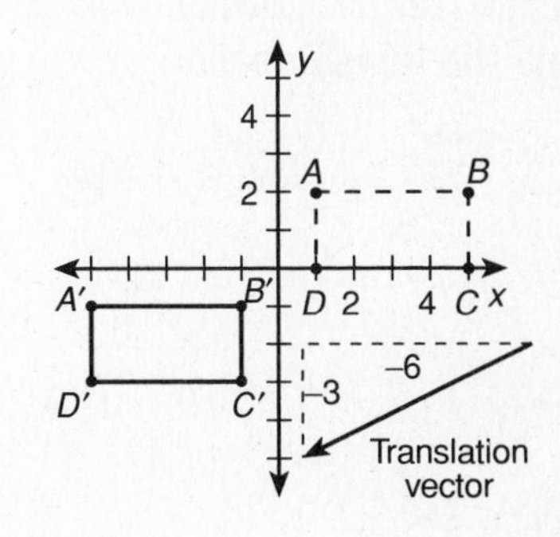

Name ______________________________

d. In the figure at the right, $\triangle ABC$ is mapped onto $\triangle RST$ by an isometry. Find the length of $\overline{RT}$, the measure of $\angle R$, and the measure of $\angle S$.

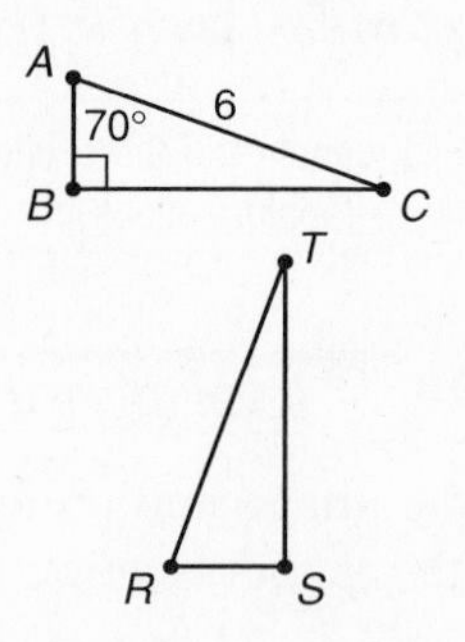

The triangles are congruent because the mapping is an isometry. Therefore, $RT = AC = 6, m\angle R = m\angle A = 70°$, and $m\angle S = m\angle B = 90°$.

e. You are designing stationery with a decorative border in the stencil pattern shown below.

If each stencil pattern measures $1\frac{1}{4}$ inches from end to end and your stationery measures $6\frac{1}{2}$ inches wide, how would you place the patterns if you wished to use 5 patterns?

By placing 5 patterns end to end, you would use $6\frac{1}{4}$ inches of space. The remaining $\frac{1}{4}$ inch would allow $\frac{1}{8}$ inch of space on each end.

Guidelines:

- The operation that maps (or moves) a preimage onto an image is called a *transformation*.
- Reflections, rotations, and translations are *rigid* transformations.

EXERCISES

In Exercises 1–3, decide whether the transformation is an isometry. If it is, name the transformation.

1. 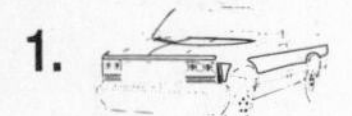 

2. 

3. 

Reteach

Chapter 7

Name ______________________________

What you should learn:

7.2	How to use properties of reflections and relate reflections and line symmetry

Correlation to Pupil's Textbook:

Mid-Chapter Self-Test (p. 344) Exercises 4, 5, 9, 11–14, 19

Chapter Test (p. 369) Exercises 11, 15–17

Examples *Using Reflections and Relating Reflections and Line Symmetry*

a. Prove Theorem 7.1; A reflection is an isometry; for the case when P is on line ℓ, and Q is not on ℓ.

Given: A reflection in ℓ maps P onto P' and Q onto Q'. P is on line ℓ, Q is not on line ℓ.

Prove: $PQ = P'Q'$

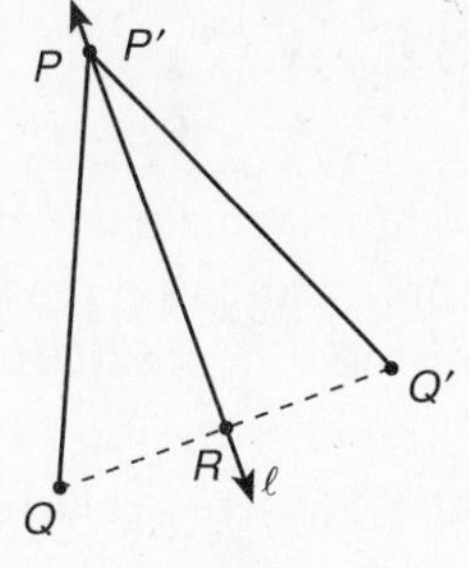

Proof:

Statements	*Reasons*
1. A reflection in ℓ maps P onto P' and Q onto Q'.	**1.** Given
2. P is on line ℓ.	**2.** Given
3. $P = P'$	**3.** Def. of a reflection
4. Q is not on line ℓ.	**4.** Given
5. ℓ is the perpendicular bisector of $\overline{QQ'}$.	**5.** Def. of a reflection
6. $\angle PRQ$ and $\angle P'RQ'$ are right angles.	**6.** $\perp$ lines intersect to form rt. $\angle$s.
7. $\angle PRQ \cong \angle P'RQ'$	**7.** All rt. angles are congruent.
8. $\overline{PR} \cong \overline{P'R}$	**8.** Reflexive Prop. of Congruence
9. $\overline{QR} \cong \overline{Q'R}$	**9.** Def. of segment bisector
10. $\triangle PRQ \cong \triangle P'RQ'$	**10.** SAS Congruence Postulate
11. $\overline{PQ} \cong \overline{P'Q'}$	**11.** CPCTC
12. $PQ = P'Q'$	**12.** Def. of congruence

b. Show that any point in the coordinate plane can be reflected about the y-axis.

Every point $Q(x, y)$ is mapped onto the point $Q'(-x, y)$, where the y-axis is the perpendicular bisector of $\overline{QQ'}$.

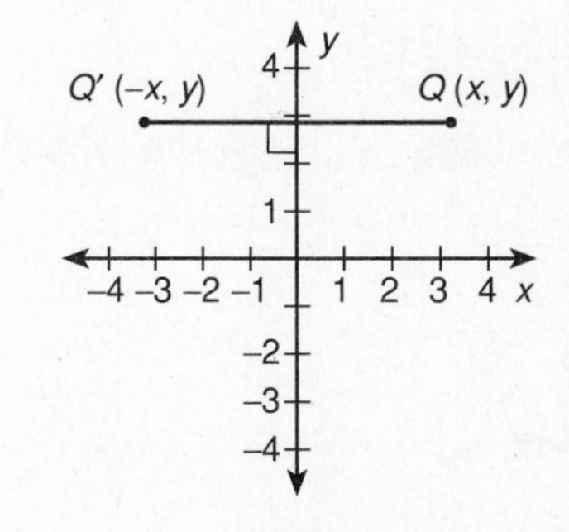

c. A figure in the plane has a line of symmetry if the figure can be mapped onto itself by a reflection. The figure at the right has how many lines of symmetry?

This figure has four lines of symmetry.

Name ____________________

d. The figure at the right has how many lines of symmetry?
This figure has one line of symmetry.

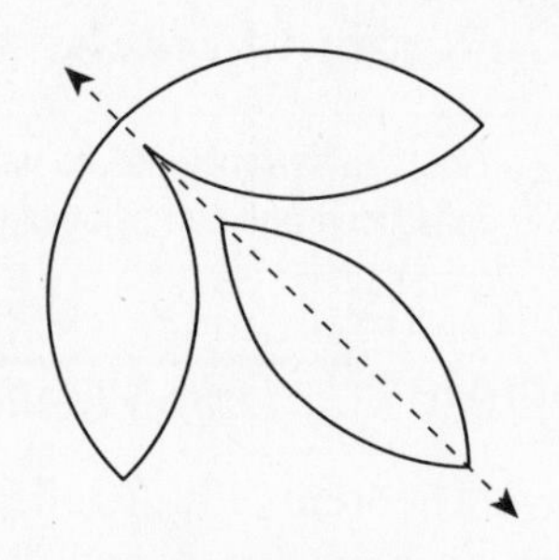

e. The kaleidoscope image shown at the right is formed with two V-shaped mirrors. Name the regular polygon and find the angle between the mirrors; the figure has how many lines of symmetry?
The image is a regular 16-gon. The angle between the mirrors is $360 \div 16$ or $22.5°$. There are 8 lines of symmetry.

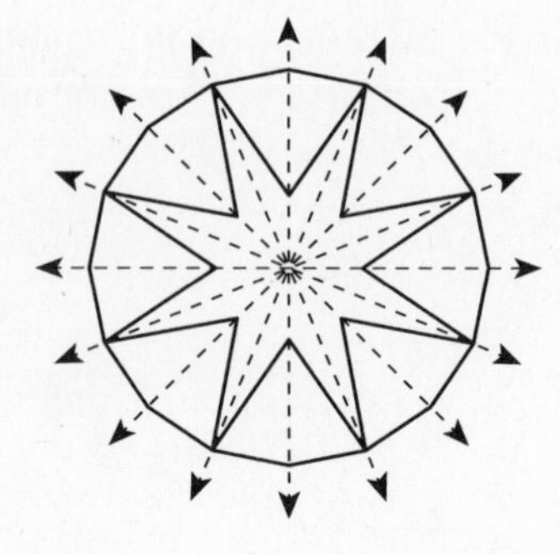

Guidelines:

- A reflection in a line ℓ is a transformation that maps every point P in the plane to a point P', so that the following is true.
  1. If P is not on ℓ, then ℓ is the perpendicular bisector of $\overline{PP'}$.
  2. If P is on ℓ, then $P = P'$.
- A reflection is an isometry.
- A figure in the plane has a line of symmetry if the figure can be mapped onto itself by a reflection.

EXERCISES

In Exercises 1–3, draw the reflection of the parallelogram in line ℓ.

1.

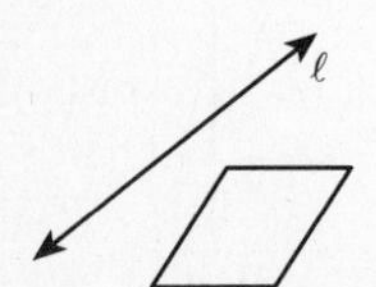

2.

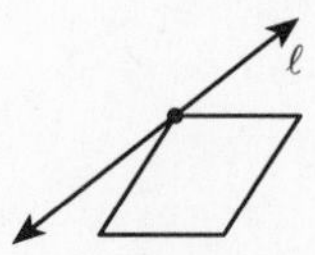

3.

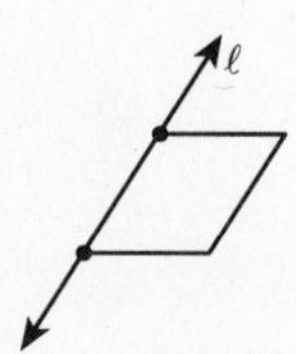

Reteach
Chapter 7

Name ________________________________

What you should learn:

7.3 How to use properties of rotations and relate rotations and rotational symmetry

Correlation to Pupil's Textbook:

Mid-Chapter Self-Test (p. 344) Exercises 3, 6, 10, 11–14, 15–18, 20

Chapter Test (p. 369) Exercises 5, 6, 11, 12, 15–17

Examples *Using Rotations and Relating Rotations and Rotational Symmetry*

a. In the figure at the right, the mapping of $\triangle ABC$ onto $\triangle A'B'C'$ is a rotation of 60°. Describe the rotation as clockwise or counterclockwise. Find segments equal in length to $\overline{PB}$ and $\overline{AB}$.

The rotation is counterclockwise. By the definition of a rotation about a point P, $PB = PB'$. Since a rotation is an isometry, $AB = A'B'$.

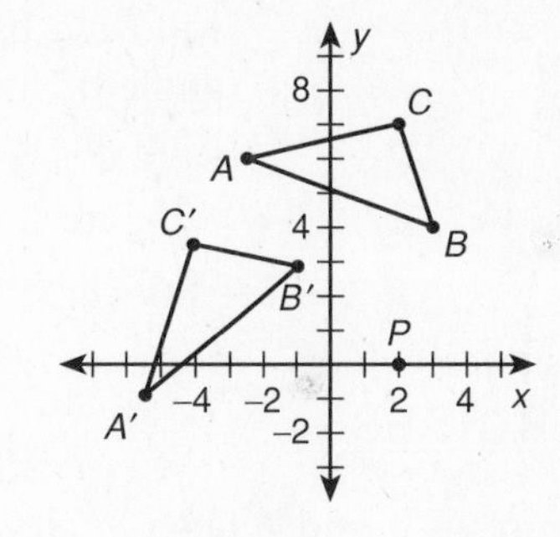

b. Lines m and n intersect at point O, as shown in the figure at the right. Rectangle $ABCD$ is reflected in m, then reflected in n. Describe the transformation that results.

The result is a rotation about point O. If $x°$ is the measure of the angle between m and n, then the angle of rotation is $2x°$, counterclockwise.

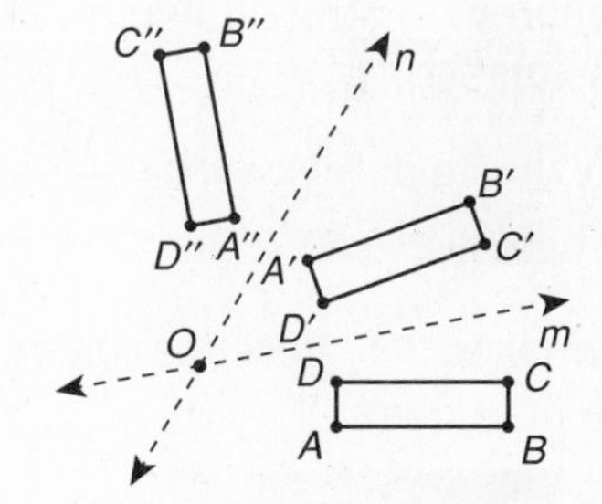

c. Describe any rotations (of 180° or less) that will map the figure shown at the right onto itself.

This figure has rotational symmetry about its center. It will be mapped onto itself if it is rotated either clockwise or counterclockwise 90° or 180°.

d. Find the image of the segment or triangle.

	Answers:
1. 90° counterclockwise rotation of $\overline{CB}$ about O.	$\overline{AH}$
2. 180° rotation of $\overline{CB}$ about O.	$\overline{GF}$
3. 90° clockwise rotation of $\triangle FEK$ about O.	$\triangle HGL$
4. 180° rotation of $\triangle CDJ$ about O.	$\triangle GHL$

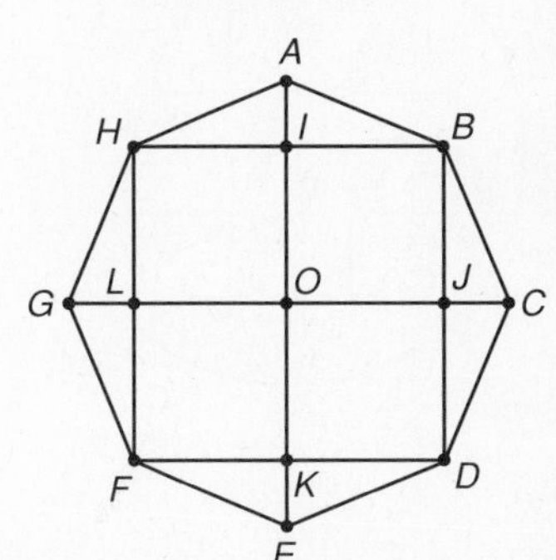

Reteach
Chapter 7

Name ______________________________

Guidelines:

- A rotation about a point O through x degrees is a transformation that maps every point P in the plane to a point P', so that the following properties are true.
  1. If P is not point O, then $PO = P'O$ and $m\angle POP' = x^\circ$.
  2. If P is point O, then $P = P'$.
- If lines ℓ and m intersect at point O, then a reflection in ℓ followed by a reflection in m is a rotation about point O. The angle of rotation is $2x^\circ$, where x° is the measure of the acute or right angle between ℓ and m.
- A figure in the plane has rotational symmetry if the figure can be mapped onto itself by a rotation of 180° or less.

EXERCISES

In Exercises 1 and 2, lines m and n intersect at point O. Consider a reflection of $\triangle XYZ$ in line m, then reflected in line n. Make a sketch of each situation.

1. If the angle of rotation about O is 54°, what is the acute angle between m and n?

2. If the angle between m and n is 54°, what is the angle of rotation about O?

In Exercises 3–6, describe any rotations (of 180° or less) that will map each figure onto itself.

3.

4.

5.

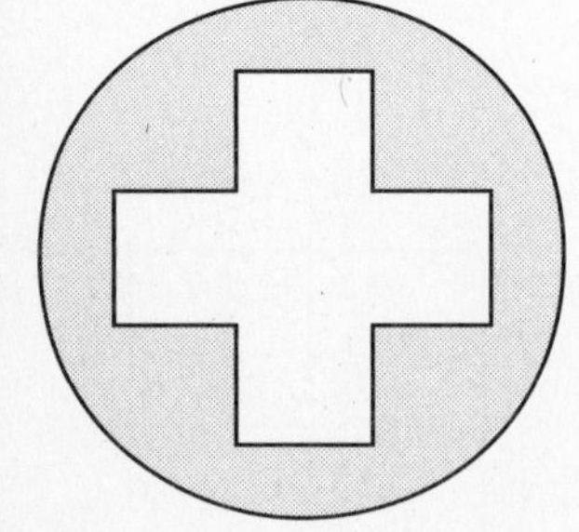

6.

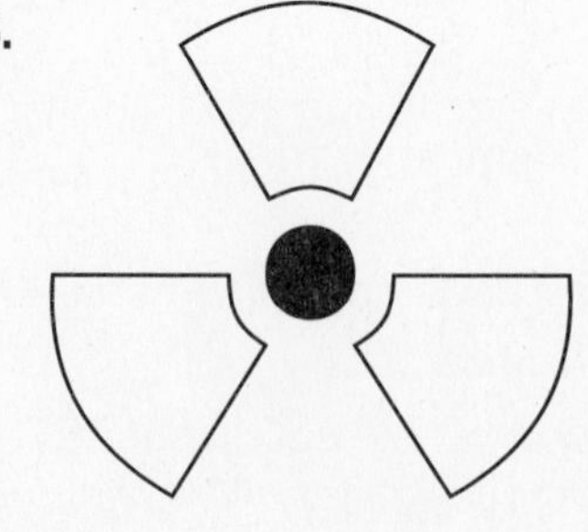

Reteach

Chapter 7

Name ____________________

What you should learn:

7.4	How to use properties of translations and to solve real-life problems

Correlation to Pupil's Textbook:

Chapter Test (p. 369)
Exercises 8, 10

Examples — *Using Properties of Translations and Solving Real-Life Problems*

a. A translation by a vector $\overrightarrow{AA'}$ is a transformation that maps every point P in the plane to a point P' so that $PP' = AA'$. Also, $\overline{PP'} \parallel \overline{AA'}$ or $\overline{PP'}$ is collinear with $\overline{AA'}$. Use the figure at the right to describe the vertical and horizontal change for the translation vector.

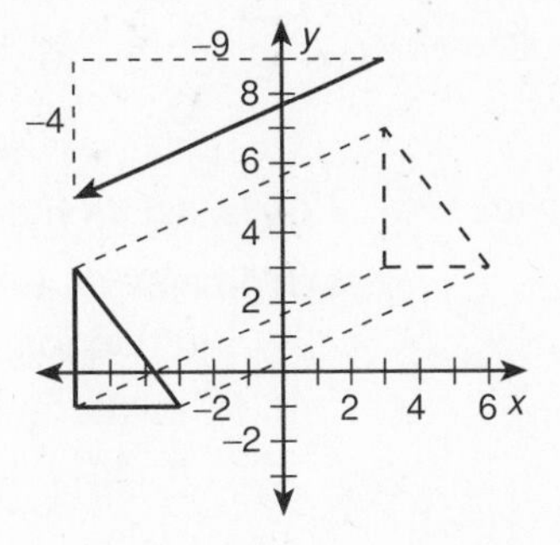

The points in a coordinate plane are moved 4 units down and 9 units to the left.

b. Use a straightedge and dot paper to translate parallelogram $ABCD$ by the vector $\vec{v} = \langle -2, -4 \rangle$.

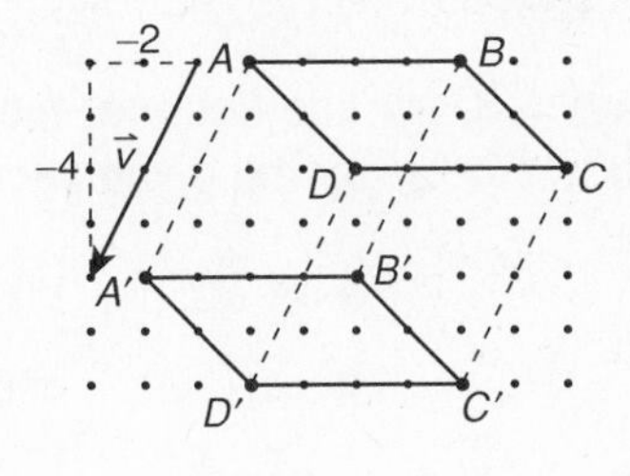

Draw the vector $\langle -2, -4 \rangle$ four times, once with its initial point at A, once with its initial point at B, once with its initial point at C, and once with its initial point at D.

Label the terminal points A', B', C', and D'. Use the straightedge to draw parallelogram $A'B'C'D'$.

c. If lines ℓ and m are parallel, then a reflection in line ℓ followed by a reflection in line m is a translation. In the figure at the right, which triangle is a translation of $\triangle XYZ$? What is the length of $\overline{PP''}$ if the distance between lines ℓ and m is d?

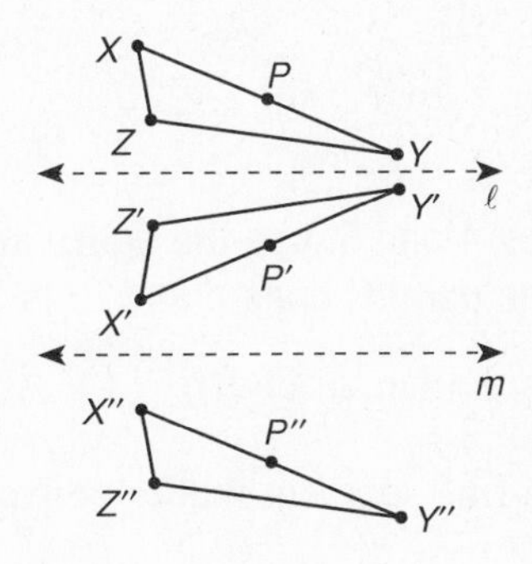

Triangle $\triangle X''Y''Z''$ is a translation of $\triangle XYZ$. The length of $\overline{PP''}$ is $2d$.

Name ____________________________

d. Use the four-note measure at the right to write a musical phrase that has one translation.

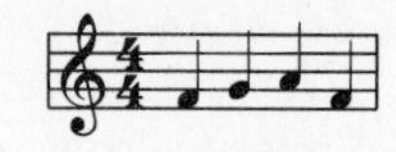

In the following musical phrase, the second measure is a translation of the first measure.

Guidelines:

- A translation is an isometry.
- If lines ℓ and m are parallel, then a reflection in line ℓ followed by a reflection in line m is a translation.
  If P'' is the image of P, then $\overline{PP''}$ is perpendicular to ℓ and $PP'' = 2d$, where d is the distance between ℓ and m.

EXERCISES

In Exercises 1–3, use the figure at the right to match the translation of $\triangle XYZ$ to $\triangle X'Y'Z'$ by the given vector.

1. $\langle 1, 0\rangle$ **a.** $X'(2, 3)$, $Y'(0, 1)$, $Z'(6, 1)$

2. $\langle -2, -1\rangle$ **b.** $X'(5, 4)$, $Y'(3, 2)$, $Z'(9, 2)$

3. $\langle 0, 3\rangle$ **c.** $X'(4, 7)$, $Y'(2, 5)$, $Z'(8, 5)$

In Exercises 4 and 5, use the figure at the right. In the figure, the distance between the parallel lines ℓ and m is 3.

4. What is the length of $\overline{BB''}$?

5. Name a line segment that is perpendicular to line ℓ.

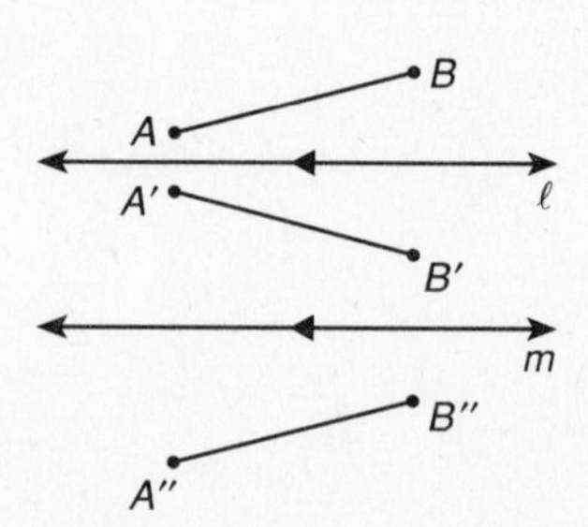

Reteach

Chapter 7

Name ______________________________

What you should learn:

7.5	How to use properties of glide reflections and compositions of transformations

Correlation to Pupil's Textbook:

Chapter Test (p. 369)

Exercises 7, 13, 14

Examples *Using Properties of Glide Reflections and Compositions*

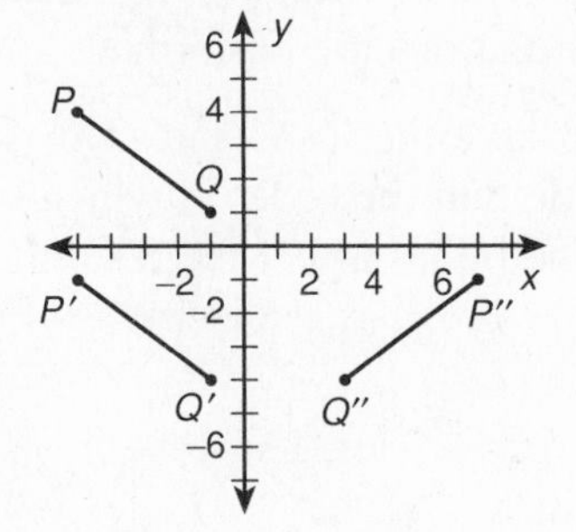

a. In the figure at the right, $\overline{PQ}$ is the preimage of a glide reflection. A glide reflection is a transformation that consists of a translation by a vector, followed by a reflection in a line that is parallel to the vector.

Name the segment that is a translation of $\overline{PQ}$ and name the vector of translation.

$\overline{P'Q'}$ is the segment that is a translation of $\overline{PQ}$ and $\vec{v} = \langle 0, -5 \rangle$.

Name the segment that is a reflection of $\overline{P'Q'}$ and name the line of reflection.

$\overline{P''Q''}$ is the segment that is a reflection of $\overline{P'Q'}$ and the line of reflection is $x = 1$.

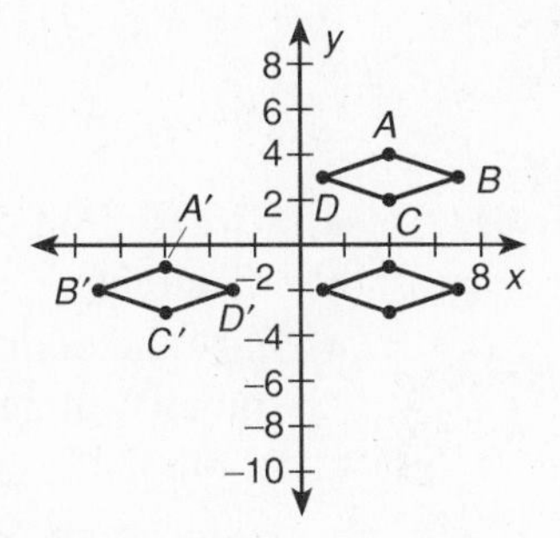

b. Find the image of the glide reflection of $ABCD$ in which $\vec{u} = \langle 0, -5 \rangle$ and ℓ is the line $x = -1$.

Begin by translating the rhombus $ABCD$ 5 units down. Then, reflect the result to obtain rhombus $A'B'C'D'$. The vertices of the image are $A'(-6, -1)$, $B'(-9, -2)$, $C'(-6, -3)$ and $D'(-3, -2)$.

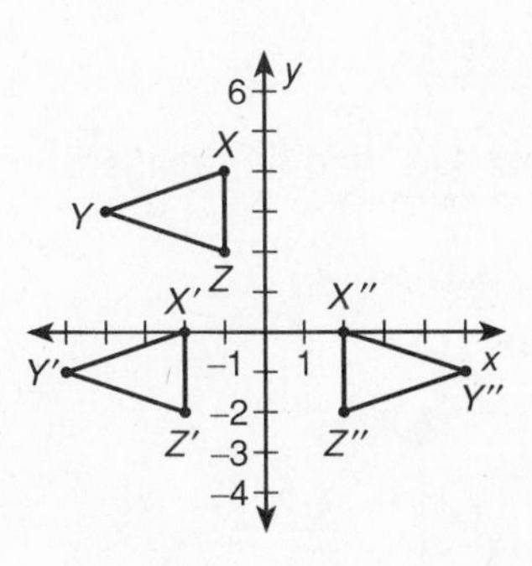

c. When two or more transformations are combined to produce a single transformation, the result is called a composition of the transformations.

In the figure at the right, $\triangle XYZ$ is translated by the vector $\vec{v} = \langle -1, -4 \rangle$ then reflected in the y-axis. Describe the composition image of $\triangle XYZ$.

The composition image is $\triangle X''Y''Z''$ whose vertices are $X''(2, 0)$, $Y''(5, -1)$, and $Z''(2, -2)$.

Name ______________________________

d. In Figure 1 at the right, $\overline{OX}$ is the preimage of a composition of isometries. First $\overline{OX}$ is rotated 90° clockwise, then reflected in the y-axis. $\overline{OX'}$ is the result of this composition.

In Figure 2 at the right, preimage $\overline{OX}$ is first reflected in the y-axis, then rotated 90° clockwise. $\overline{OX''}$ is the result of this composition.

What conclusion can you make about the order in which two transformations are performed?

You observe that $\overline{OX'}$ and $\overline{OX''}$ are different images and conclude that the order in which two transformations are performed does affect the resulting composition.

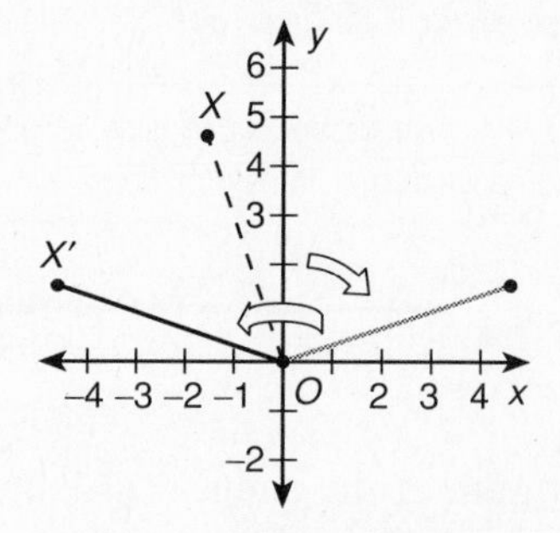

Figure 1

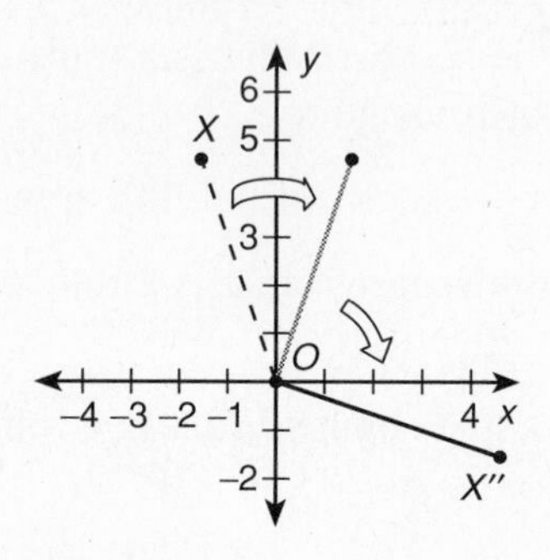

Figure 2

Guidelines:

- A translation can be thought of as a composition of two reflections.
- Since the composition of two (or more) isometries is an isometry, glide reflections are isometries.
- The order in which two transformations are performed affects the resulting composition.

EXERCISES

In Exercises 1–3, find the image of the indicated glide reflection.

1. $\vec{v} = \langle 3, 0 \rangle$
ℓ: x-axis

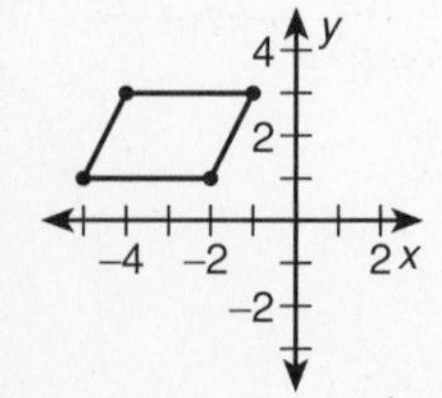

2. $\vec{v} = \langle 0, 2 \rangle$
ℓ: y-axis

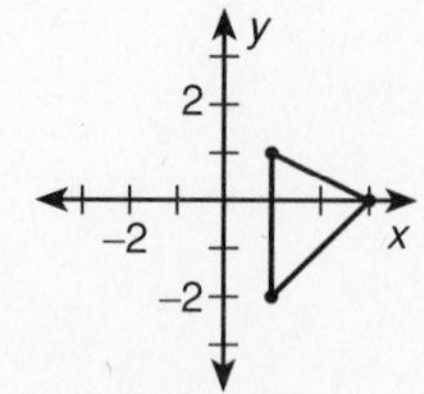

3. $\vec{v} = \langle 2, 2 \rangle$
ℓ: $y = x$

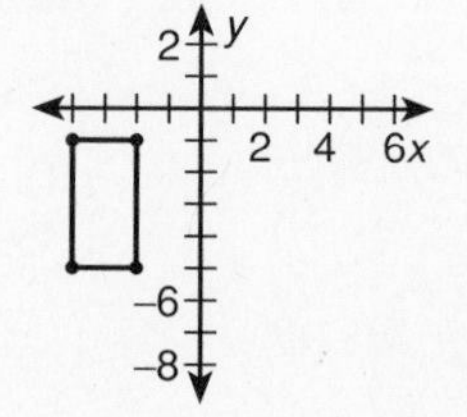

Reteach

Chapter 7

Name ________________________

What you should learn:

7.6	How to use transformations to classify frieze patterns and use frieze patterns in real life

Correlation to Pupil's Textbook:

Chapter Test (p. 369)
Exercises 18–20

Examples — *Classifying Frieze Patterns and Using Frieze Patterns in Real Life*

a. A frieze pattern or strip pattern is a pattern that extends indefinitely to the left and right in such a way that the pattern can be mapped onto itself by a transformation.

Describe the transformation(s) that map(s) each frieze pattern shown at the right onto itself.

1. horizontal translation or 180° rotation
2. horizontal translation or horizontal glide reflection
3. horizontal translation only
4. horizontal translation or reflection about a vertical line

1. 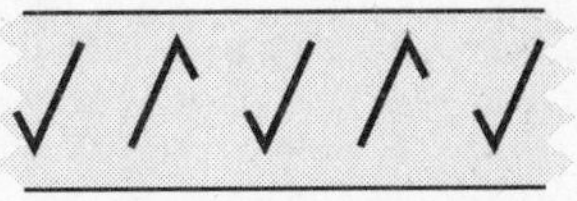
2. 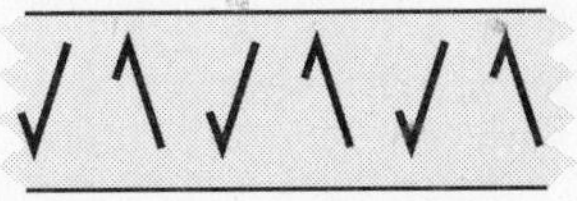
3. 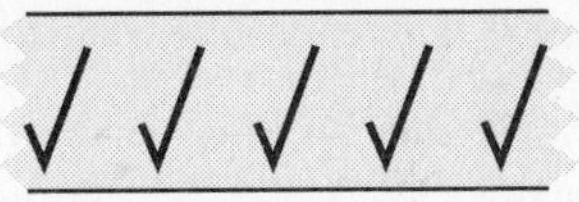
4.

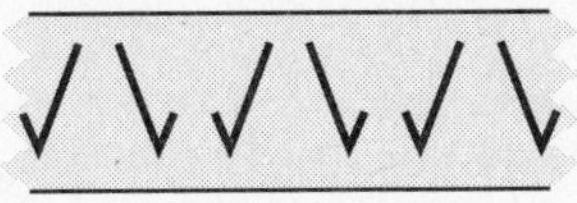

b. Every frieze pattern can be classified into one of seven categories. Classify each frieze pattern shown at the right.

T	Translation only
TR	Translation and 180° rotation
TV	Translation and vertical line reflection
TG	Translation and glide reflection
THG	Translation, horizontal line reflection, and glide reflection
TRVG	Translation, 180° rotation, vertical line reflection, and glide reflection
TRHVG	Translation, 180° rotation, horizontal line reflection, vertical line reflection, and glide reflection

1. 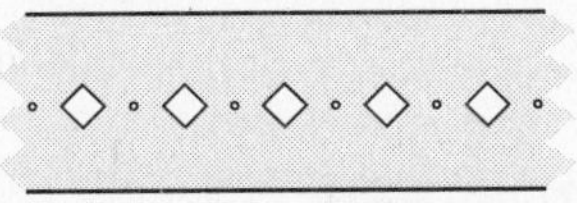
Answer: TRHVG

2. 
Answer: TRVG

3. 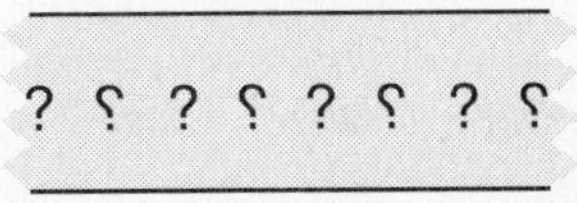
Answer: TV

Reteach
Chapter 7

Name ______________________

c. An example of Corinthian Greek architecture is shown at the right.

Describe the horizontal frieze pattern located below the cornice and above the architrave.

The frieze pattern can be mapped onto itself by a translation, a 180° rotation, a reflection about a horizontal line, a reflection about a vertical line, and a glide reflection.

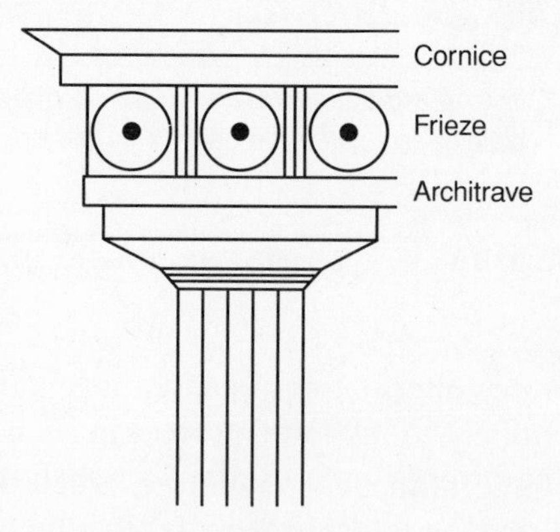

d. One row of a quilt is shown at the right. Classify the horizontal frieze pattern of the quilt.

The frieze pattern is a horizontal translation.

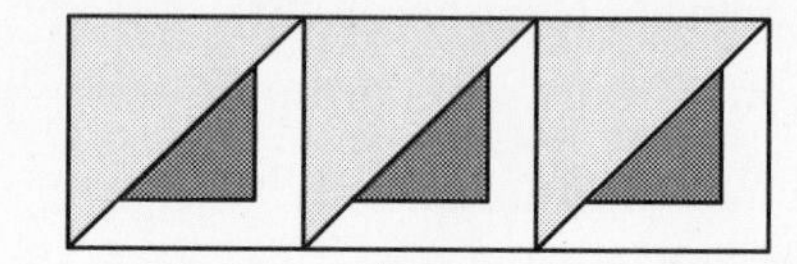

Guidelines:

- To classify a frieze pattern into one of seven categories, you can use a tree diagram.
- A frieze pattern cannot be a translation and horizontal line reflection because a pattern that has a horizontal line reflection must also have a glide reflection.

EXERCISES

In Exercises 1–4, classify the frieze pattern.

1. 

2. 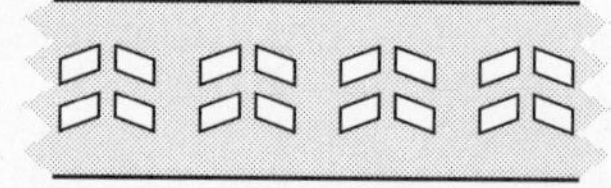

3. 

4. 

Reteach

Chapter 8

Name ____________________

What you should learn:

8.1	How to compute the ratio of two numbers and use proportions to solve problems

Correlation to Pupil's Textbook:

Mid-Chapter Self-Test (p. 393) Exercises 1–10

Chapter Test (p. 423) Exercises 1–4

Examples — *Computing Ratios and Using Proportions*

a. A hospital clinic has 14 medical employees and 6 clerical employees. The following ratios compare two quantities that are measured in the same units.

The ratio of medical employees to clerical employees is $\dfrac{14 \text{ employees}}{6 \text{ employees}} = \dfrac{7}{3}$.

The ratio of clerical employees to total employees is $\dfrac{6 \text{ employees}}{20 \text{ employees}} = \dfrac{3}{10}$.

The ratio of total employees to medical employees is $\dfrac{20 \text{ employees}}{14 \text{ employees}} = \dfrac{10}{7}$.

b. A proportion is an equation that equates two ratios, such as $\frac{x}{3} = \frac{4}{5}$. You can solve this proportion by using the cross product property, $x \cdot 5 = 3 \cdot 4$. Since $5x = 12$, $x = \frac{12}{5}$. Check this by substituting into the original proportion, $\dfrac{\frac{12}{5}}{3} = \dfrac{12}{15} = \dfrac{4}{5}$. It checks.

Guidelines:

- If a and b are two quantities that are measured in the same units, then the *ratio* of a and b is $\frac{a}{b}$ or $a{:}b$.
- In the proportion $\frac{a}{b} = \frac{c}{d}$, a and d are the *extremes* of the proportion. The *means* of the proportion are b and c.
- If two ratios are equal, then their reciprocals are equal.

EXERCISES

In Exercises 1–3, solve the proportion.

1. $\dfrac{3}{4} = \dfrac{y}{7}$

2. $\dfrac{c-1}{4} = \dfrac{c}{6}$

3. $\dfrac{2}{a} = \dfrac{5}{a+2}$

4. The ratio $CD{:}DE$ is 5:6. Use a proportion to solve for x.

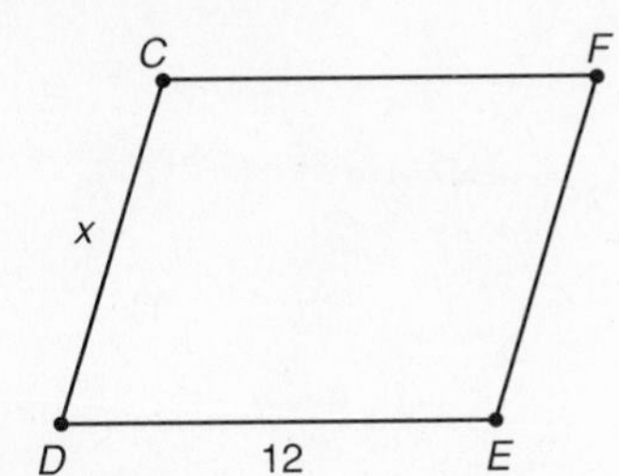

5. The ratio $PQ{:}QR$ is 4:3. Use a proportion to solve for x.

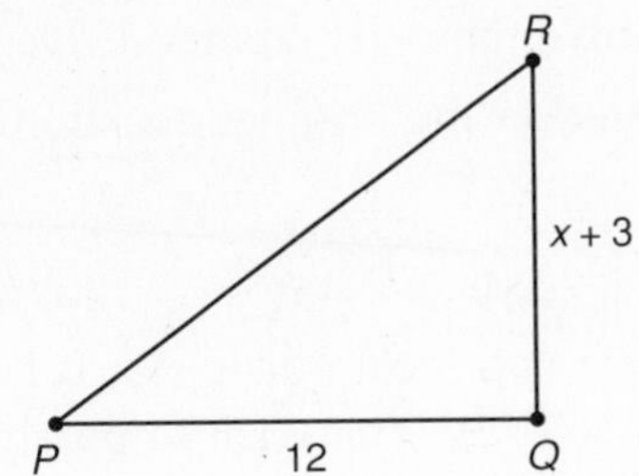

Reteach
Chapter 8

Name ______________________________

What you should learn:

8.2	How to use properties of proportions and use a problem-solving plan

Correlation to Pupil's Textbook:

Mid-Chapter Self-Test (p. 393) Exercises 9, 10, 17–20

Chapter Test (p. 423) Exercises 1–4

Examples *Using Properties of Proportions and Using a Problem-Solving Plan*

a. Given the proportion $\frac{x}{2} = \frac{5}{8}$, you can interchange the means, resulting in the equivalent proportion $\frac{x}{5} = \frac{2}{8}$.

b. Given the proportion $\frac{x}{2} = \frac{5}{8}$, you can add the denominators 2 and 8 to the numerators x and 5, resulting in the equivalent proportion

$$\frac{x+2}{2} = \frac{5+8}{8}.$$

c. Your family's home is valued at $90,000 and the real estate taxes are $2160. If your family remodels the garage and builds a new deck, the home will be valued at $98,000. Assuming that the rate of tax stays the same, how much should your family expect to pay in real estate taxes after the remodeling?

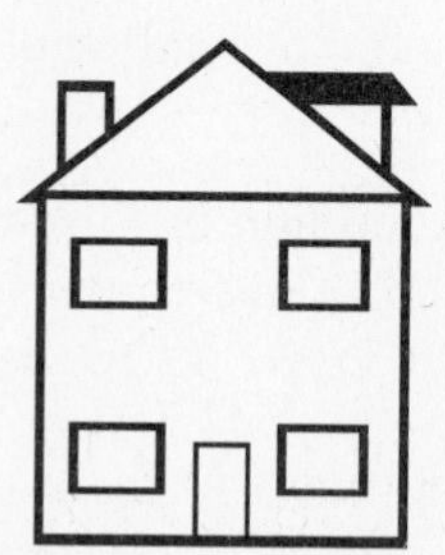

Verbal model: $\frac{\text{Current taxes}}{\text{Current value of home}} = \frac{\text{Taxes after remodeling}}{\text{Value of home after remodeling}}$

Label: Taxes after Remodeling $= x$ (dollars)

Equation: $\frac{2160}{90{,}000} = \frac{x}{98{,}000}$

$\frac{2160}{90{,}000} \cdot 98{,}000 = \frac{x}{98{,}000} \cdot 98{,}000$ *Multiply both sides by* 98,000.

$2352 = x$ *Simplify.*

Your family should expect to pay $2352 after the remodeling.

Guidelines: To use a problem-solving plan for problems involving proportions:

- Write a verbal model.
- Assign labels.
- Write an algebraic model.
- Solve the algebraic model.
- Answer the original question and check the answer.

EXERCISE

Compact discs are on sale at a local music store. During the sale, you can purchase three CD's for $27.99. You want to purchase 5 of your favorite CD's. How much will you pay for them?

Reteach
Chapter 8

Name ______________________

What you should learn:

8.3	How to identify similar polygons and use similar polygons to solve problems

Correlation to Pupil's Textbook:

Mid-Chapter Self-Test (p. 393)
Exercises 9–16

Chapter Test (p. 423)
Exercises 6–8, 10

Examples *Identifying and Using Similar Polygons*

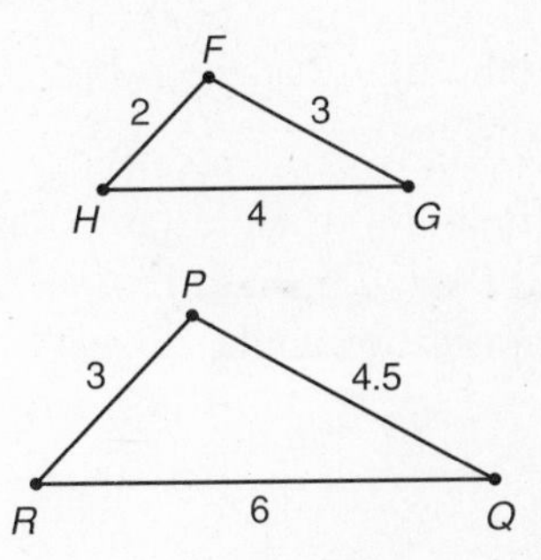

a. In the figure at the right, $\triangle FGH \sim \triangle PQR$. This means that corresponding angles are congruent and the lengths of corresponding sides are proportional.

Angles: $\angle F \cong \angle P$, $\angle G \cong \angle Q$, $\angle H \cong \angle R$

Sides: $\frac{FG}{PQ} = \frac{3}{4.5} = \frac{2}{3}$, $\frac{GH}{QR} = \frac{4}{6} = \frac{2}{3}$, $\frac{HF}{RP} = \frac{2}{3}$

The scale factor of $\triangle FGH$ to $\triangle PQR$ is $\frac{2}{3}$ or 2:3.

b. Find the perimeters of the triangles shown above. The perimeter of $\triangle FGH$ is 9 units and the perimeter of $\triangle PQR$ is 13.5 units. Since the polygons are similar, the ratio of their perimeters $\frac{9}{13.5} = \frac{2}{3}$, the ratio of the corresponding sides.

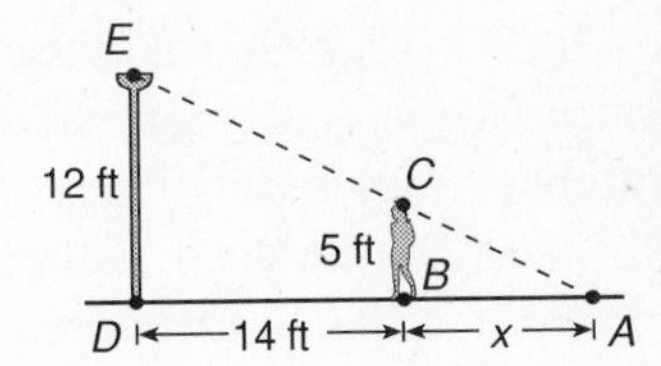

c. Find the length of the shadow of a girl who is 5 feet tall and is standing 14 feet from a floodlight that is mounted 12 feet high. Use similar triangles $\triangle ABC \sim \triangle ADE$ to find x, the length of the shadow. Since lengths of corresponding sides of similar triangles are proportional, $\frac{AB}{AD} = \frac{CB}{ED}$. Substituting, $\frac{x}{14+x} = \frac{5}{12}$. Solving the proportion gives the solution $x = 10$. The shadow is 10 feet long.

Guidelines:

- Two polygons are similar if their corresponding angles are congruent and the lengths of their corresponding sides are proportional.
- If two polygons are similar, then the ratio of their perimeters is equal to the ratio of their corresponding sides.

EXERCISES

In Exercises 1–3, determine if the polygons are similar. Explain your reasoning.

1.

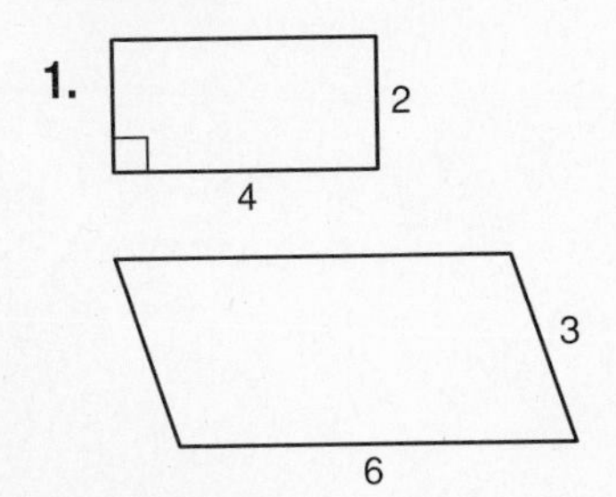

2.

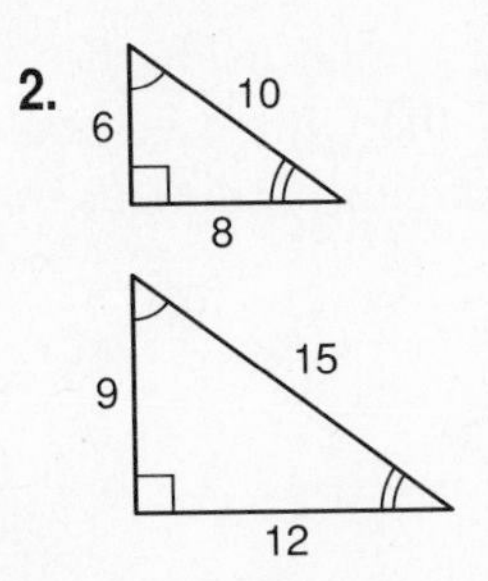

3.

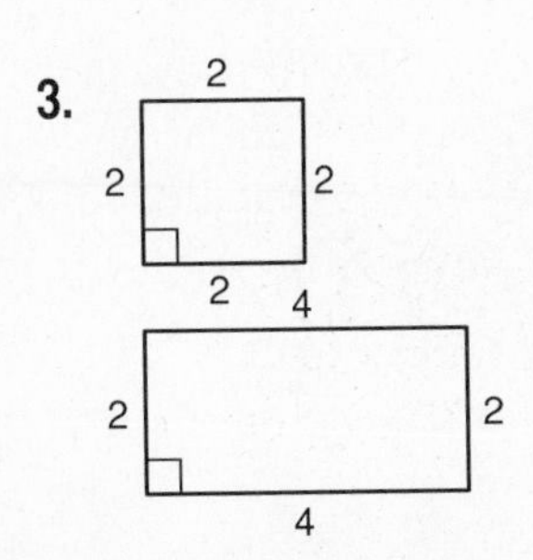

Reteach
Chapter 8

Name ____________________

What you should learn:

8.4	How to identify similar triangles and use similar triangles in coordinate geometry

Correlation to Pupil's Textbook:

Chapter Test (p. 423)
Exercises 5, 9, 11–14

Examples *Identifying Similar Triangles and Using Similarity*

a. In the figure at the right, $\triangle ABC \sim \triangle EFG$. Find $m\angle G$, $m\angle E$, and $m\angle B$.

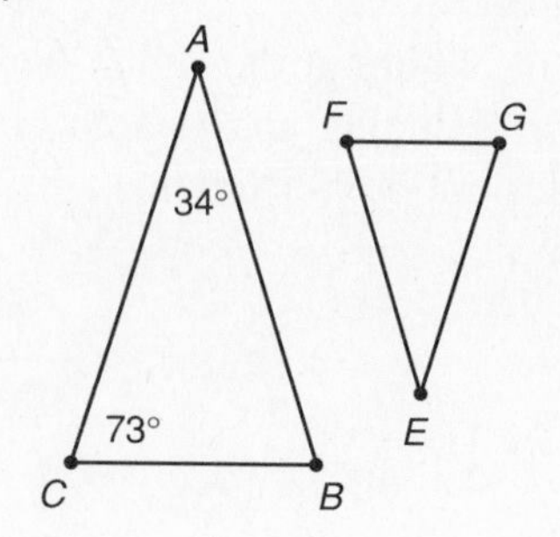

Given the triangles are similar, $\angle C \cong \angle G$ and $\angle A \cong \angle E$. Since $m\angle C = 73°$ and $m\angle A = 34°$, $m\angle G = 73°$ and $m\angle E = 34°$. The measure of angle $\angle B$ is

$$180 - m\angle A - m\angle C = 180° - 34° - 73°$$
$$= 73°.$$

b. In the figure at the right, parallel lines ℓ_1 and ℓ_2 are cut by transversals t_1 and t_2. Show that $\triangle ABC \sim \triangle DEC$.

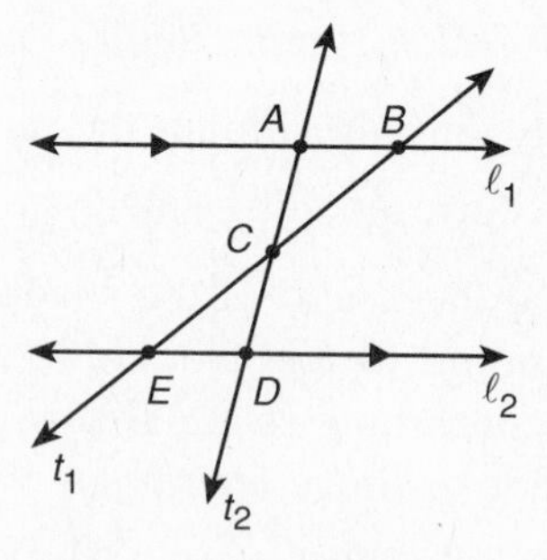

By the Alternate Interior Angles Theorem and transversal t_1, $\angle ABC \cong \angle DEC$. Likewise, using transversal t_2, $\angle BAC \cong \angle EDC$. Since two angles of one triangle are congruent to two angles of another triangle, then the two triangles are similar (Angle-Angle Similarity Postulate). (Note: Could also use $\angle ACB \cong \angle DCE$ by the Vertical Angles Theorem.)

c. Describe a plan for a proof.

Given: $\angle RQP \cong \angle RSQ$

Prove: $\triangle RQP \sim \triangle RSQ$

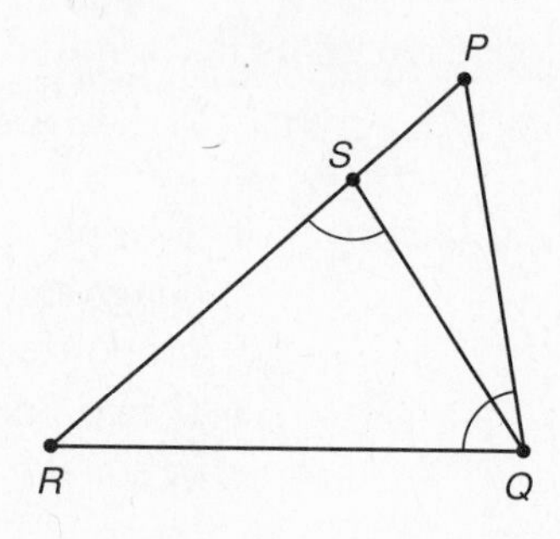

Since $\angle R \cong \angle R$ by the Reflexive Property of Congruence, triangles $\triangle RQP$ and $\triangle RSQ$ have two pairs of congruent angles. By the Angle-Angle Similarity Postulate, the triangles are similar.

Guidelines:

- You can verify that any two points on a nonvertical line can be used to calculate the slope of the line, using ratios of corresponding sides of similar triangles.
- To show that two triangles are similar using the Angle-Angle Similarity Postulate, you must show that two angles of one triangle are congruent to two angles of another triangle.

Reteach
Chapter 8

Name ______________________________

EXERCISES

1. Use the figure at the right to calculate the slope of $\overleftrightarrow{AC}$ using points A and B, B and C, and A and C. Name two similar triangles in the figure.

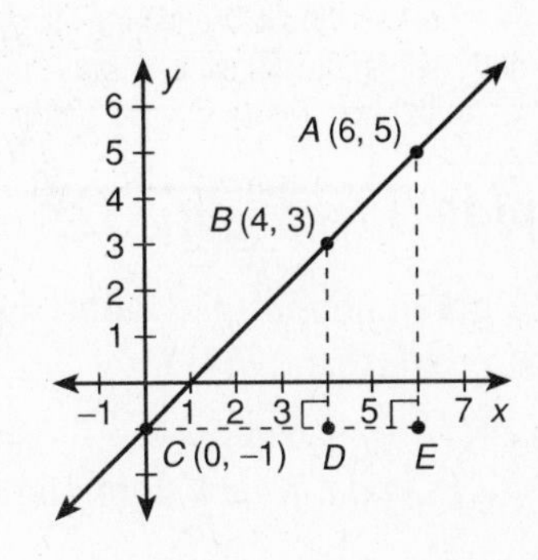

2. Complete the proof.
Given: $\overline{WY} \parallel \overline{VZ}$
Prove: $\triangle XWY \sim \triangle XVZ$

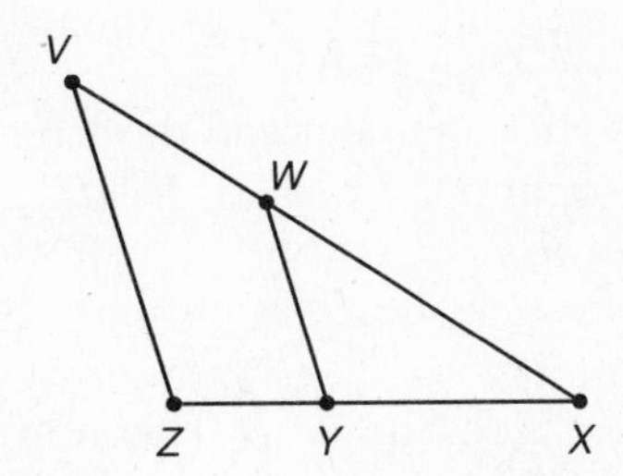

Proof:

Statements	*Reasons*
1. $\overline{WY} \parallel \overline{VZ}$	1. Given
2. $\angle XWY \cong \angle XVZ$	2. ?
3. $\angle XYW \cong \angle XZV$	3. ?
4. $\triangle XWY \sim \triangle XVZ$	4. ?

In Exercises 3–5, determine whether the triangles are similar. Explain your reasoning.

3. Is $\triangle ABC \sim \triangle ADE$?

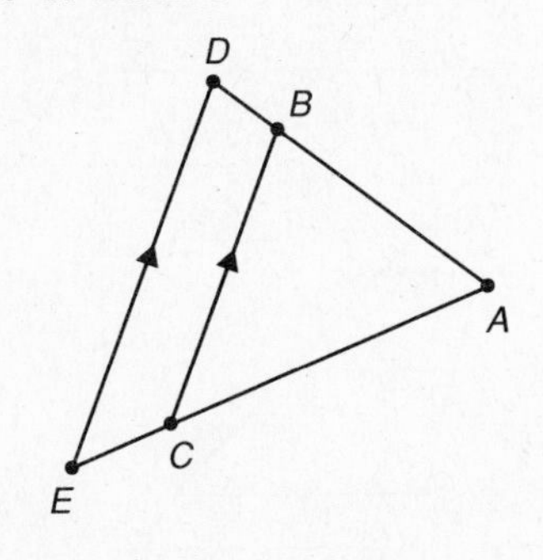

4. Is $\triangle ABC \sim \triangle DEF$?

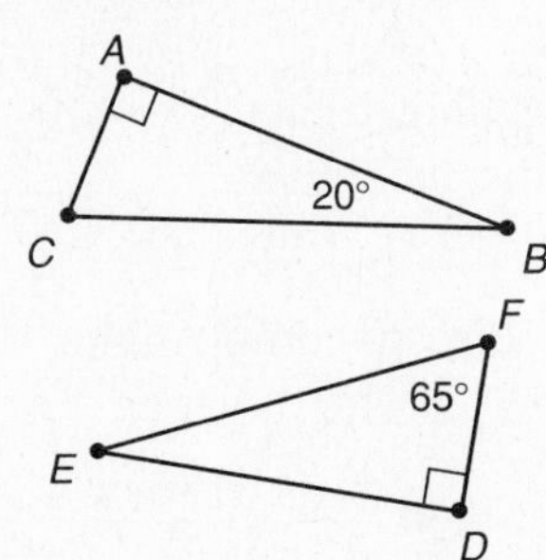

5. Is $\triangle ABC \sim \triangle DCE$?

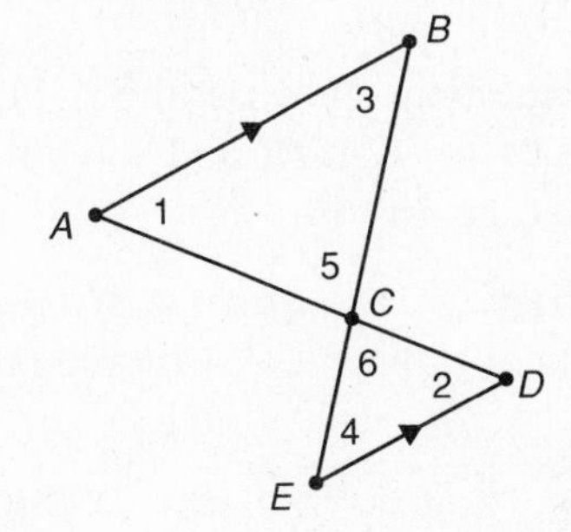

Reteach
Chapter 8

Name ______________________

What you should learn:

8.5	How to use the SSS and SAS Similarity Theorems and use similar triangles to solve real-life problems

Correlation to Pupil's Textbook:

Chapter Test (p. 423)
Exercises 17, 18

Examples *Using the SSS and SAS Similarity Theorems*

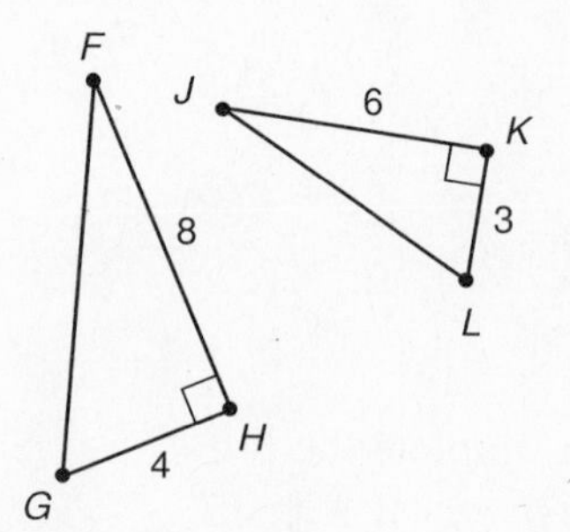

a. For the given conditions, name the postulate or theorem that can be used to prove that $\triangle FGH \sim \triangle JLK$.

Given: $FH = 8$, $JK = 6$, $GH = 4$, $LK = 3$,
$\angle H$ and $\angle K$ are right angles.

Since $\frac{FH}{JK} = \frac{8}{6} = \frac{4}{3}$ and $\frac{GH}{LK} = \frac{4}{3}$, then $\frac{FH}{JK} = \frac{GH}{LK}$.

The right angles are congruent and the lengths of the sides including the right angles are proportional. $\triangle FGH \sim \triangle JLK$ by the SAS Similarity Theorem.

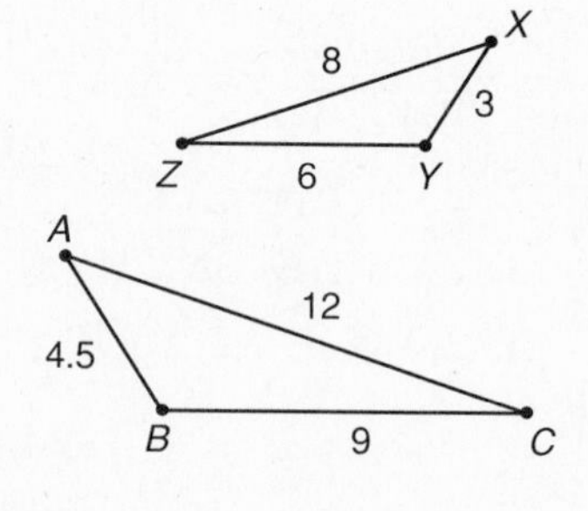

b. Use the Side-Side-Side Similarity Theorem to show $\triangle ABC \sim \triangle XYZ$.

If corresponding sides of two triangles are proportional, then the two triangles are similar. Ratios of side lengths of $\triangle ABC$ and $\triangle XYZ$ are as follows.

Shortest sides: $\frac{AB}{XY} = \frac{4.5}{3} = \frac{3}{2}$

Longest sides: $\frac{AC}{XZ} = \frac{12}{8} = \frac{3}{2}$

Remaining sides: $\frac{BC}{YZ} = \frac{9}{6} = \frac{3}{2}$

Because the ratios are equal, the triangles are similar.

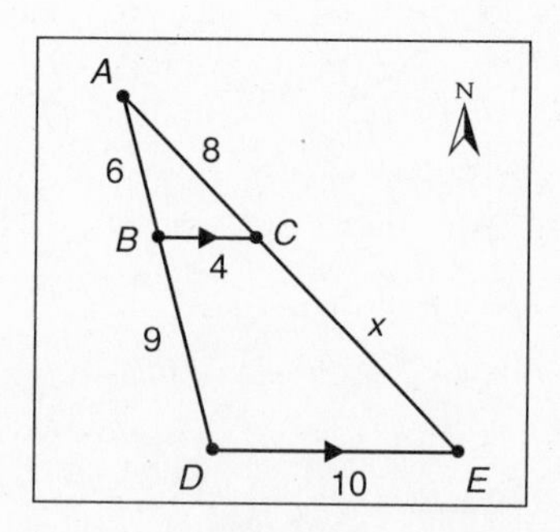

c. On the map shown at the right, $\overline{BC} \parallel \overline{DE}$. Show that $\triangle ABC \sim \triangle ADE$ and solve for CE, the distance from city C to city E.

If $\overline{BC} \parallel \overline{DE}$, then $\angle ABC \cong \angle ADE$ (Corresponding Angles Postulate). $\angle A \cong \angle A$ by reflexivity.

Since $\frac{AB}{AD} = \frac{6}{15} = \frac{2}{5}$ and $\frac{BC}{DE} = \frac{4}{10} = \frac{2}{5}$, then $\frac{AB}{AD} = \frac{BC}{DE}$.

By SAS Similarity Theorem, $\triangle ABC \sim \triangle ADE$. Let $CE = x$.
Solve the proportion

$$\frac{AC}{AE} = \frac{8}{8+x} = \frac{2}{5}.$$

The solution $x = 12$ is the distance from city C to city E.

Reteach
Chapter 8

Name ____________________

Guidelines:

- If corresponding sides of two triangles are proportional, then the two triangles are similar (SSS Similarity Theorem).
- If an angle of one triangle is congruent to an angle of a second triangle and the lengths of the sides including these angles are proportional, then the triangles are similar (SAS Similarity Theorem).

EXERCISES

1. In the figure below, $\triangle FGH \sim \triangle RST$. Find RT and ST.

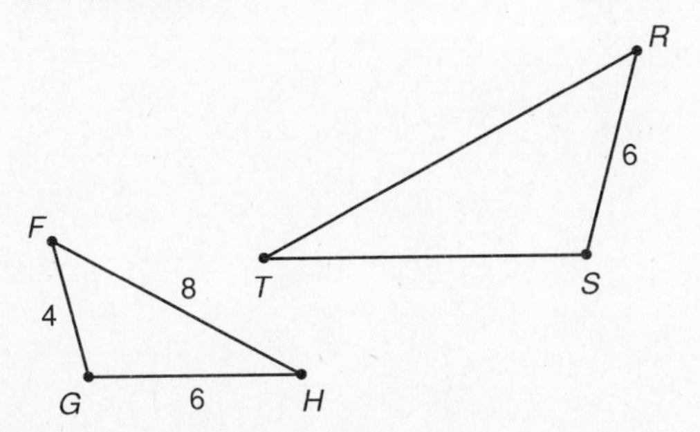

2. In the figure below, $\triangle OPN \sim \triangle OLM$. Find LM and NM.

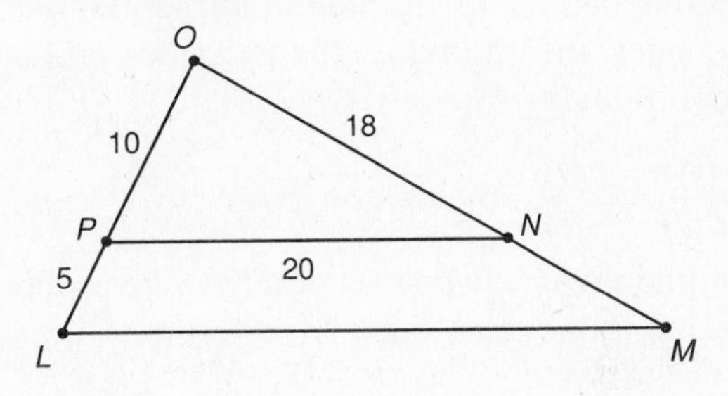

In Exercises 3–5, state the postulate or theorem that can be used to prove that the two triangles are similar.

3.

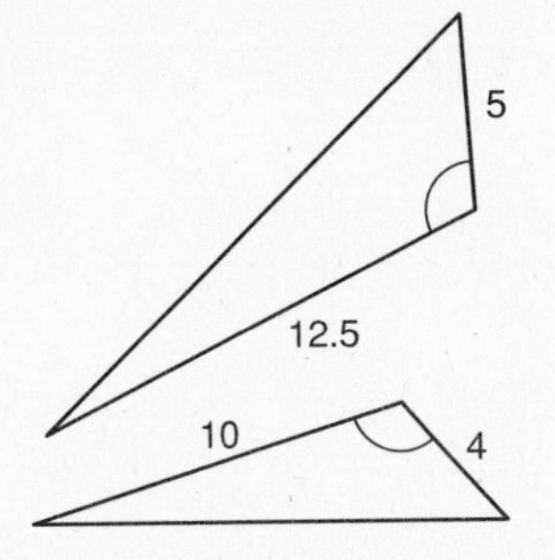

4.

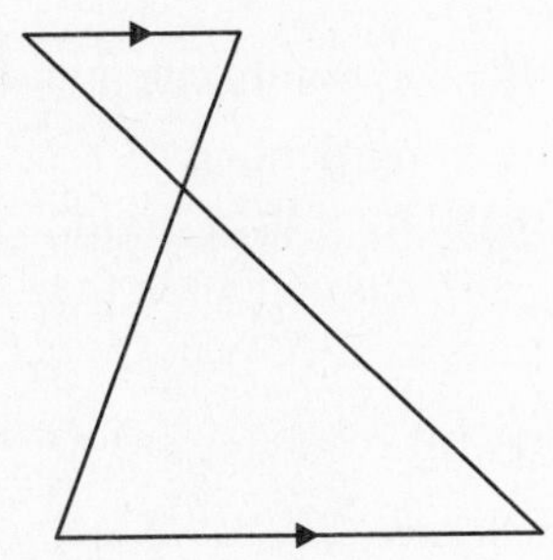

5.

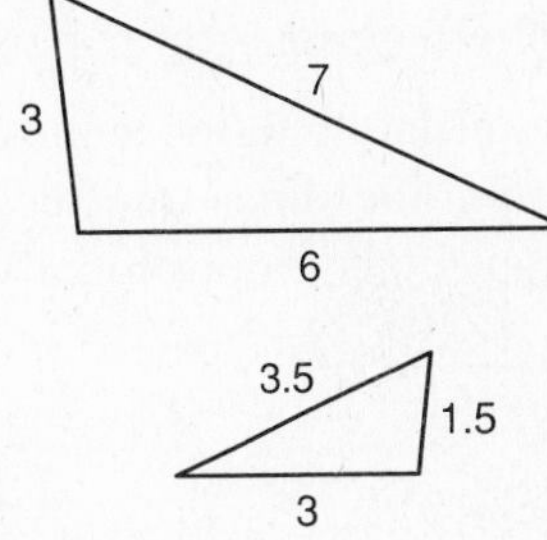

6. Complete the proof.

Given: Isosceles triangles $\triangle ABC, \triangle DEF$

$\angle C \cong \angle F$, $\frac{AB}{DE} = \frac{BC}{EF}$

Prove: $\triangle ABC \sim \triangle DEF$

Statements	*Reasons*
1. $\triangle ABC, \triangle DEF$ are isosceles triangles.	**1.** Given
2. $\angle B \cong \angle C, \angle E \cong \angle F$	**2.** [?]
3. $\angle C \cong \angle F$	**3.** Given
4. $\angle B \cong \angle E$	**4.** [?]
5. $\frac{AB}{DE} = \frac{BC}{EF}$	**5.** Given
6. $\triangle ABC \sim \triangle DEF$	**6.** [?]

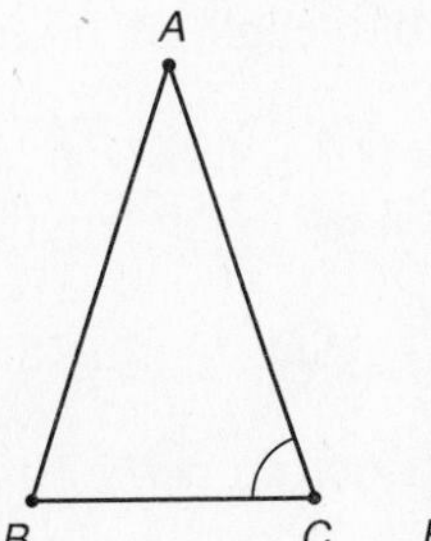

Reteach

Chapter 8

Name ______________________________

What you should learn:

8.6	How to use proportionality theorems to solve problems in geometry and in real life

Correlation to Pupil's Textbook:

Chapter Test (p. 423)
Exercises 15, 16

Examples *Using Proportionality Theorems and Solving Real-Life Problems*

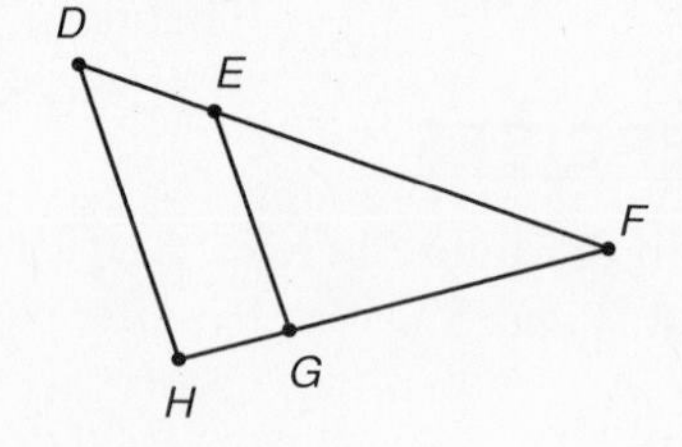

a. Use the figure at the right. State the proportionality theorem used in each case.

1. If $\overline{EG} \parallel \overline{DH}$, then $\dfrac{FE}{ED} = \dfrac{FG}{GH}$.

   If a line parallel to one side of a triangle intersects the other two sides, then it divides the two sides proportionally (Triangle Proportionality Theorem).

2. If $\dfrac{FG}{GH} = \dfrac{FE}{ED}$, then $\overline{EG} \parallel \overline{DH}$.

   If a line divides two sides of a triangle proportionally, then it is parallel to the third side (Triangle Proportionality Converse).

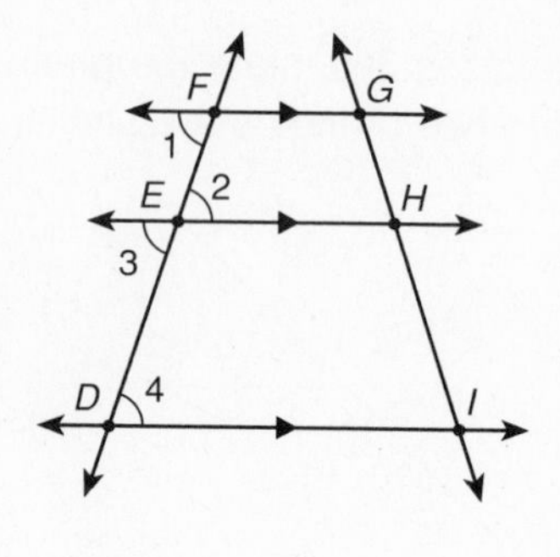

b. In the figure at the right, $\angle 1 \cong \angle 2$ and $\angle 3 \cong \angle 4$. What can you conclude about the lengths of $\overline{FE}$, $\overline{ED}$, $\overline{GH}$, and $\overline{HI}$?

Since alternate interior angles $\angle 1$ and $\angle 2$ are congruent, $\overleftrightarrow{FG} \parallel \overleftrightarrow{EH}$. Since alternate interior angles $\angle 3$ and $\angle 4$ are congruent, $\overleftrightarrow{EH} \parallel \overleftrightarrow{DI}$. If three parallel lines intersect two transversals, then they divide the transversals proportionally. Therefore, you can conclude that $\dfrac{FE}{ED} = \dfrac{GH}{HI}$.

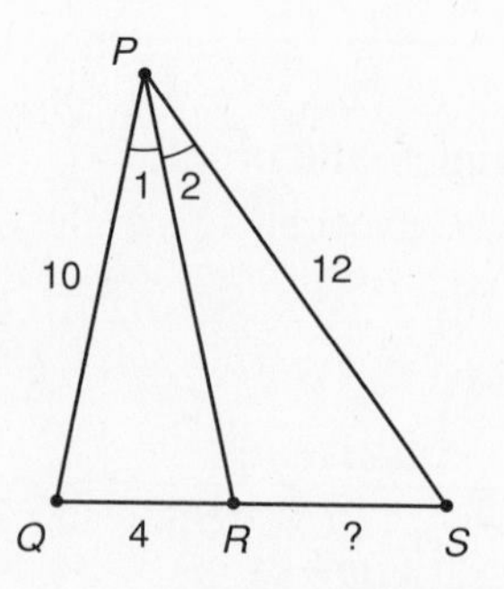

c. Use the figure at the right. If $\angle 1 \cong \angle 2$, $PQ = 10$, $PS = 12$, and $QR = 4$. What is the length of $\overline{RS}$?

Since $\overrightarrow{PR}$ bisects $\angle QPS$, then $\overrightarrow{PR}$ divides $\overline{QS}$ into segments whose lengths are proportional to the lengths of the other two sides.

That is, $\dfrac{QR}{RS} = \dfrac{PQ}{PS}$. Substituting the given values, $\dfrac{4}{RS} = \dfrac{10}{12}$.
Using the cross product property $10 \cdot RS = 48$ and $RS = 4.8$.

Name ________________________

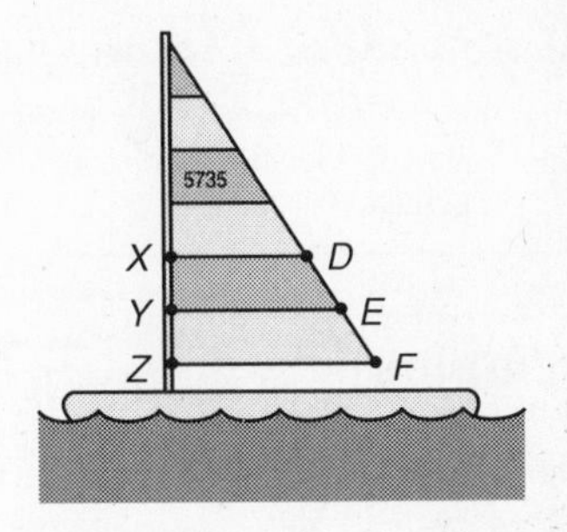

d. You are designing a new sail for a Hobie sailboat, as shown at the right. Each horizontal section of material is a different color and is evenly spaced. What can you conclude about DE and EF?

Since the horizontal sections are evenly spaced, the seams joining the section are parallel lines. $\overleftrightarrow{XD}$, $\overleftrightarrow{YE}$, and $\overleftrightarrow{ZF}$ are parallel lines intersecting two transversals, therefore, they divide the transversals proportionally. You know that $XY = YZ$; therefore, $DE = EF$.

Guidelines:

- If a line parallel to one side of a triangle intersects the other two sides, then it divides the two sides proportionally.
- If a line divides two sides of a triangle proportionally, then it is parallel to the third side.
- If three parallel lines intersect two transversals, then they divide the transversals proportionally.
- If a ray bisects an angle of a triangle, then it divides the opposite side into segments whose lengths are proportional to the lengths of the other two sides.

EXERCISES

In Exercises 1–4, use the figure at the right to complete the proportion.

1. $\dfrac{WX}{XY} = \dfrac{\boxed{?}}{\boxed{?}}$

2. $\dfrac{CB}{\boxed{?}} = \dfrac{EF}{EG}$

3. $\dfrac{\boxed{?}}{AB} = \dfrac{GD}{GF}$

4. $\dfrac{CE}{AG} = \dfrac{\boxed{?}}{DG}$

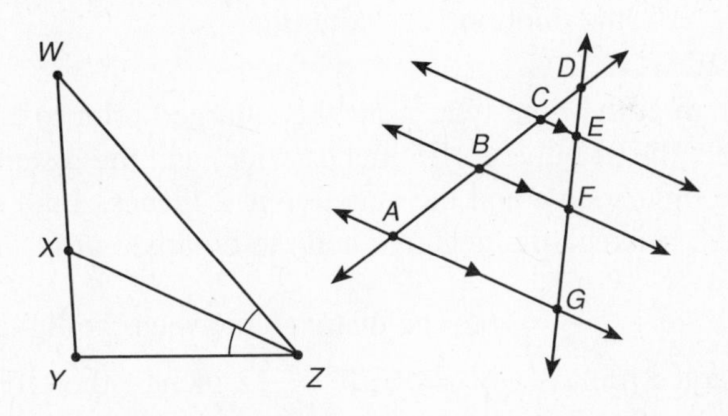

In Exercises 5–8, use the figure at the right to match the segment with its length.

a. $7\frac{1}{2}$ **b.** $2\frac{1}{2}$ **c.** 6 **d.** 2

5. SQ **6.** QP **7.** NR **8.** OQ

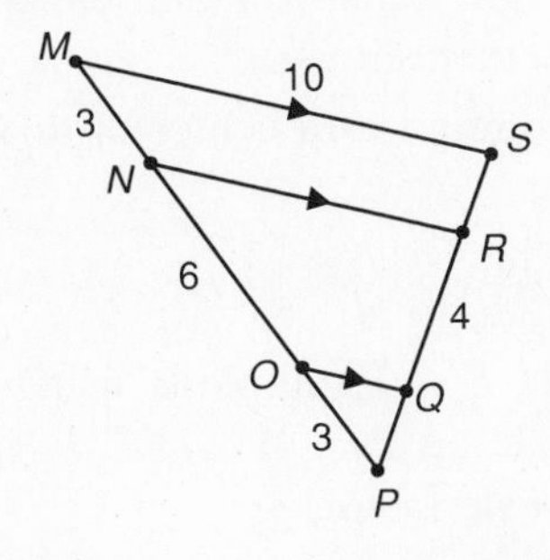

Reteach
Chapter 8

Name ______________________________

What you should learn:

8.7	How to identify dilations and use dilations to solve real-life problems

Correlation to Pupil's Textbook:

Chapter Test (p. 423)
Exercises 19, 20

Examples — *Identifying Dilations and Using Dilations in Real Life*

a. Identify the dilation, and find its scale factor.

The dilation with center C has scale factor $k = \frac{3}{2}$ because $\frac{CP'}{CP} = \frac{3}{2}$.

The dilation is an enlargement because $k > 1$.

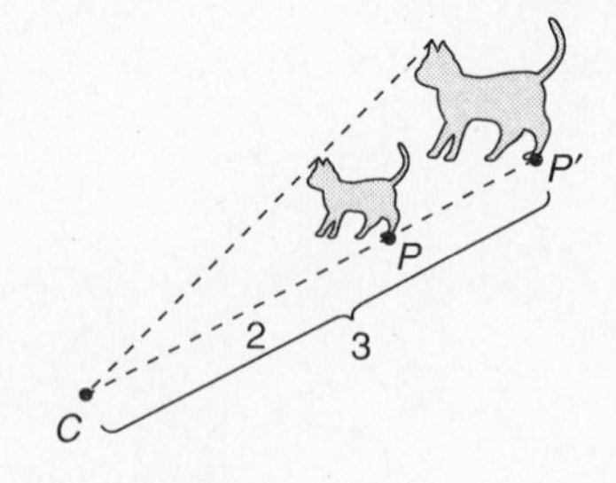

b. Identify the dilation of triangle DEF in the coordinate plane.

The center, C, of the dilation is the origin and the image of each point $P(x, y)$ is $P'(kx, ky)$. You can find the scale factor k by dividing

$$\frac{CD'}{CD} = \frac{CE'}{CE} = \frac{CF'}{CF} = \frac{1}{2}.$$

Since $0 < k < 1$, the dilation is a reduction.

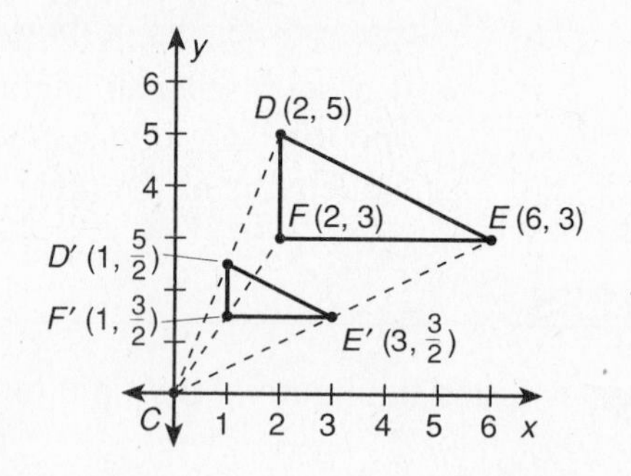

c. In the diagram at the right, you want the enlarged print to be 9 inches wide. The negative is 1.5 inches wide, and the distance between the light source and the negative is 2 inches. Find the distance, SS', between the negative and the enlarged print.

The scale factor is $\frac{9}{1.5} = 6$. The distance between the light source and the enlarged print is $CS' = 6(2) = 12$ inches; therefore, the distance SS' is $12 - 2 = 10$ inches.

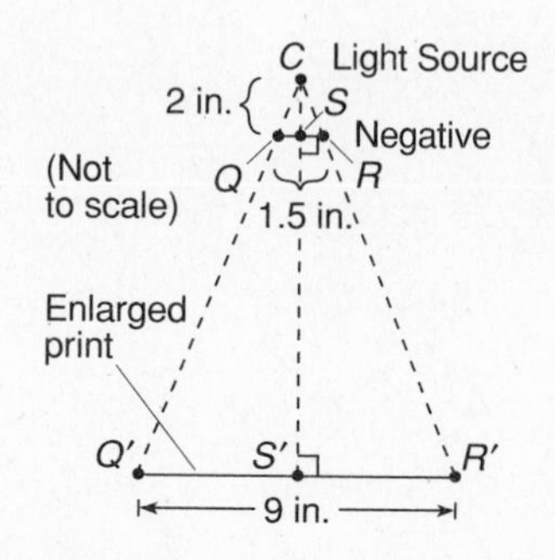

Guidelines:

- In a nonrigid transformation called a dilation, every image is similar to its preimage.
- A dilation requires a center of dilation and a scale factor.

EXERCISE

Draw a dilation of triangle $\triangle DEF$ in Example b, using the scale factor $k = 2$. Label the image triangle $\triangle XYZ$. If $\triangle D'E'F'$ is mapped onto $\triangle XYZ$ by a dilation, find the scale factor.

Reteach
Chapter 9

Name ______________________

What you should learn:

9.1	How to prove right triangles congruent and use properties of right triangles

Correlation to Pupil's Textbook:

Mid-Chapter Self-Test (p. 447)	Chapter Test (p. 473)
Exercises 1–8, 15, 16	Exercises 1–6

Examples *Proving Right Triangles Congruent Using Properties of Right Triangles*

a. In the figure at the right, $\angle MNO$ and $\angle NPO$ are right angles. Name three similar triangles.

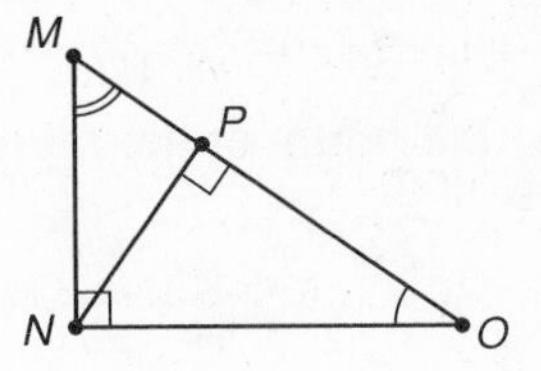

Since an altitude $\overline{PN}$ is drawn to the hypotenuse $\overline{MO}$ of right triangle $\triangle OMN$, the two triangles $\triangle ONP$ and $\triangle NMP$ are similar to the original triangle $\triangle OMN$ and to each other.

b. Using the figure in Example a, explain how the length of $\overline{PN}$ is related to the lengths of $\overline{PM}$ and $\overline{PO}$.

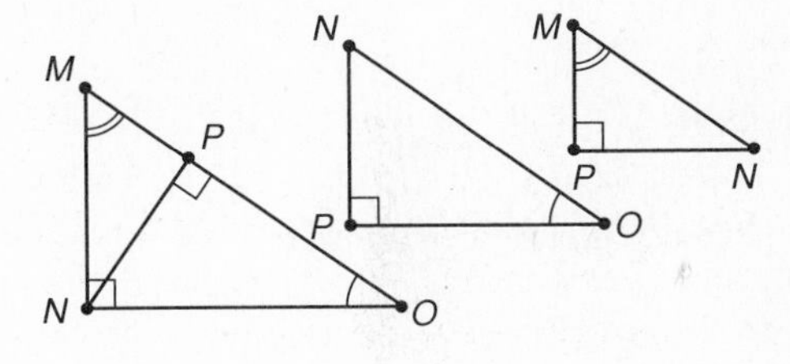

Since $\overline{PN}$ is an altitude from the right angle to the hypotenuse of right triangle $\triangle OMN$, PN is the geometric mean of PM and PO (the lengths of the segments of the hypotenuse). Therefore,

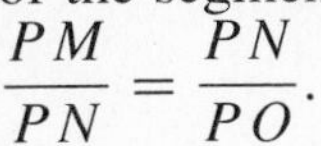

$$\frac{PM}{PN} = \frac{PN}{PO}.$$

c. In the figure at the right, $\angle EFG$ and $\angle FHG$ are right angles. Find the length of $\overline{EF}$ and $\overline{GF}$.

Each leg of right triangle $\triangle EFG$ is the geometric mean of the hypotenuse and the segment of the hypotenuse that is adjacent to the leg.

$\frac{12}{EF} = \frac{EF}{3}$	Def. of geometric mean	$\frac{12}{GF} = \frac{GF}{9}$
$(EF)^2 = 36$	Cross multiply.	$(GF)^2 = 108$
$EF = 6$	Take positive square root.	$GF = 6\sqrt{3}$

d. Quadrilateral $ABCD$ is a rectangle. Prove that $\triangle ABC \cong \triangle CDA$.

Given: Quadrilateral $ABCD$ is a rectangle.

Prove: $\triangle ABC \cong \triangle CDA$

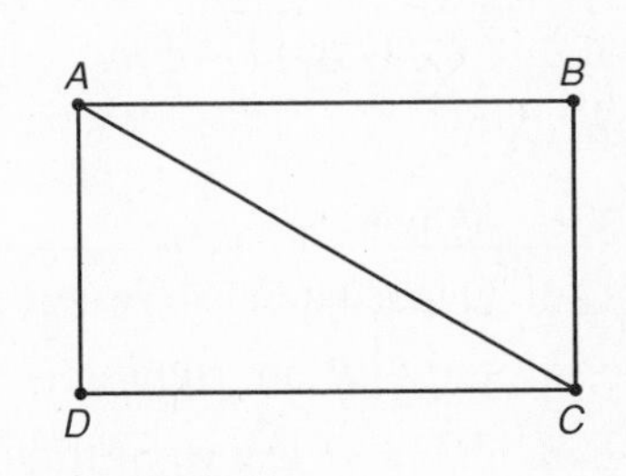

Proof:

Statements	*Reasons*
1. Quadrilateral $ABCD$ is a rectangle.	**1.** Given
2. $\angle B$, $\angle D$ are right angles.	**2.** Def. of a rectangle
3. $\triangle ABC$ is a right triangle.	**3.** Def. of right triangle
4. $\triangle CDA$ is a right triangle.	**4.** Def. of right triangle
5. Quadrilateral $ABCD$ is a parallelogram.	**5.** Def. of a rectangle
6. $\overline{AD} \cong \overline{CB}$	**6.** Property of a Parallelogram
7. $\overline{AC} \cong \overline{AC}$	**7.** Reflexive Prop. of Congruence
8. $\triangle ABC \cong \triangle CDA$	**8.** HL Congruence Theorem

Reteach

Chapter 9

Name ______________________

Guidelines:

- You can use proportions to solve for segment lengths in a right triangle with an altitude to the hypotenuse.
- You can prove that two right triangles are congruent using the Hypotenuse-Leg (HL) Congruence Theorem.

EXERCISES

In Exercises 1–5, use the figure at the right to complete each statement.

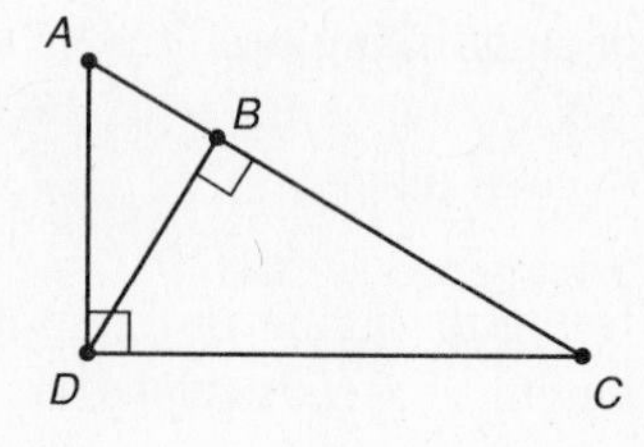

1. $\triangle ADC \sim$ [?] $\sim$ [?]

2. $\dfrac{AB}{[?]} = \dfrac{[?]}{CB}$

3. $\dfrac{AC}{AD} = \dfrac{AD}{[?]}$

4. $\dfrac{AC}{[?]} = \dfrac{[?]}{CB}$

5. If $AB = 2$ and $CB = 6$, then $BD =$ [?].

6. State a reason for each step of the proof.
Given: $\overline{AB} \perp \overline{BE}, \overline{FE} \perp \overline{BE}, \overline{AD} \cong \overline{FC}, \overline{BD} \cong \overline{EC}$
Prove: $\triangle ABD \cong \triangle FEC$

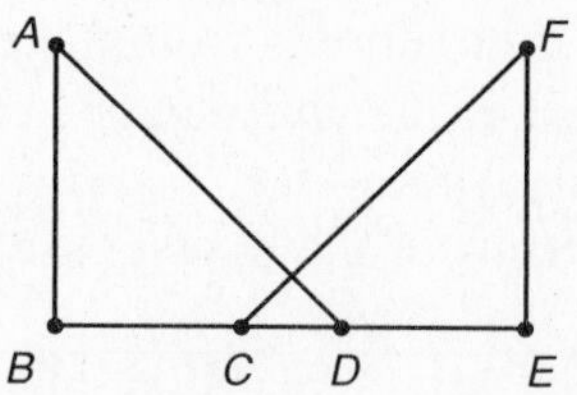

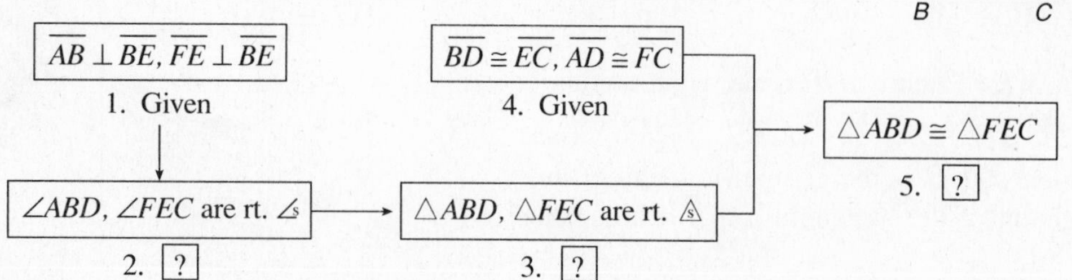

7. State a reason for each step of the proof.
Given: Quadrilateral $MNOP$ is a rectangle.
Prove: $\triangle MPO \cong \triangle PMN$
Proof:

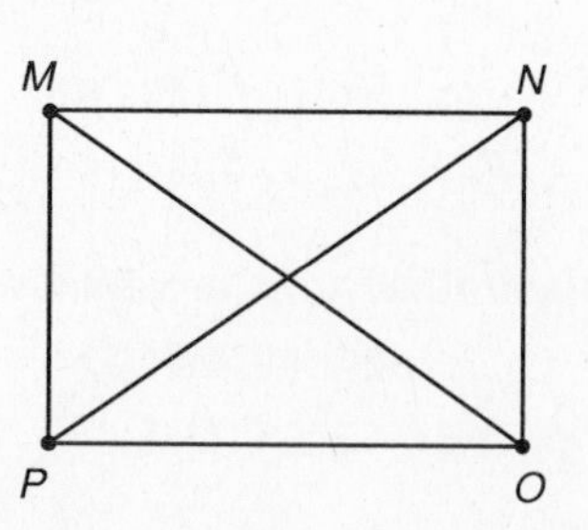

Statements	*Reasons*
1. Quadrilateral $MNOP$ is a rectangle.	**1.** [?]
2. $\angle M, \angle P$ are right angles.	**2.** [?]
3. $\triangle MPO$ is a right triangle.	**3.** [?]
4. $\triangle PMN$ is a right triangle.	**4.** [?]
5. $\overline{MO} \cong \overline{PN}$	**5.** [?]
6. $\overline{MP} \cong \overline{MP}$	**6.** [?]
7. $\triangle MPO \cong \triangle PMN$	**7.** [?]

Reteach
Chapter 9

Name ______________________________

What you should learn:

9.2	How to prove the Pythagorean Theorem and use it to solve problems

Correlation to Pupil's Textbook:

Mid-Chapter Self-Test (p. 447) Exercises 9, 14, 17–20

Chapter Test (p. 473) Exercises 1–6, 11

Examples *Proving the Pythagorean Theorem*

a. Use the figure at the right to plan a proof of the Pythagorean Theorem. Explain how the diagram can be used to prove the theorem by adding areas.

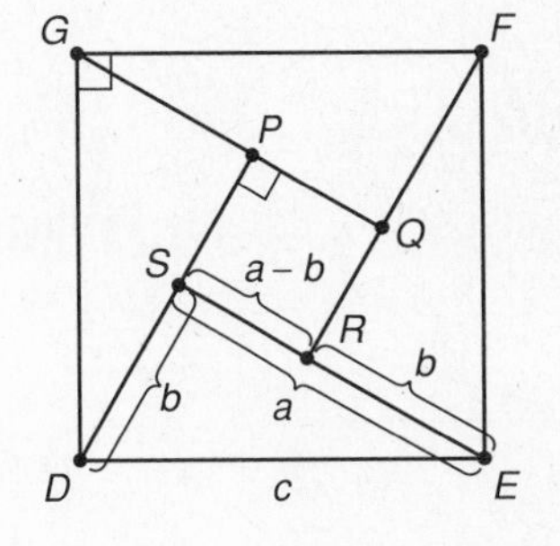

Plan for proof:

The total area of square $DEFG$ is the area of square $PQRS$ plus the areas of four congruent right triangles. The area formula for a triangle is one-half the product of the base and the height. Since each triangle in the figure has a base a and height b, the area is $\frac{1}{2}ab$. Four times $\frac{1}{2}ab = 2ab$ is the total area of the four triangles. Square $PQRS$ has side $(a - b)$ and square $DEFG$ has side c. Since the area formula for a square is side squared, the area of square $PQRS$ is $(a-b)^2$ and the area of square $DEFG$ is c^2. Adding the total area of the four triangles to the area of square $PQRS$ is the same as the area of square $DEFG$.

$$2ab + (a - b)^2 = c^2$$
$$2ab + a^2 - 2ab + b^2 = c^2$$
$$a^2 + b^2 = c^2$$

b. A 30-foot tree casts a shadow 16 feet long. Use the diagram at the right to find x, the distance from the top of the tree to the tip of the shadow.

30 ft

x

16 ft

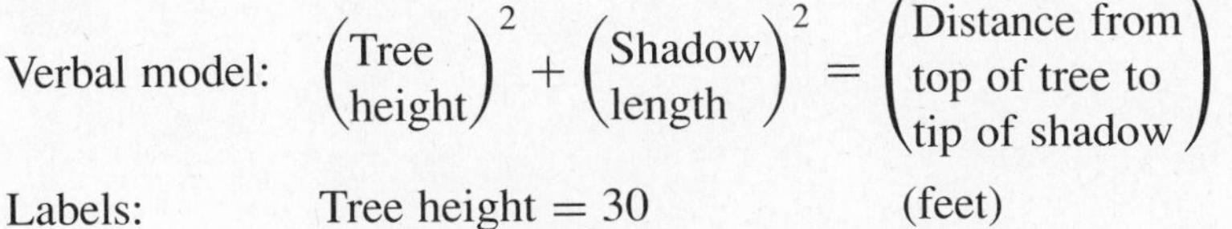

Verbal model: $\begin{pmatrix}\text{Tree}\\\text{height}\end{pmatrix}^2 + \begin{pmatrix}\text{Shadow}\\\text{length}\end{pmatrix}^2 = \begin{pmatrix}\text{Distance from}\\\text{top of tree to}\\\text{tip of shadow}\end{pmatrix}^2$

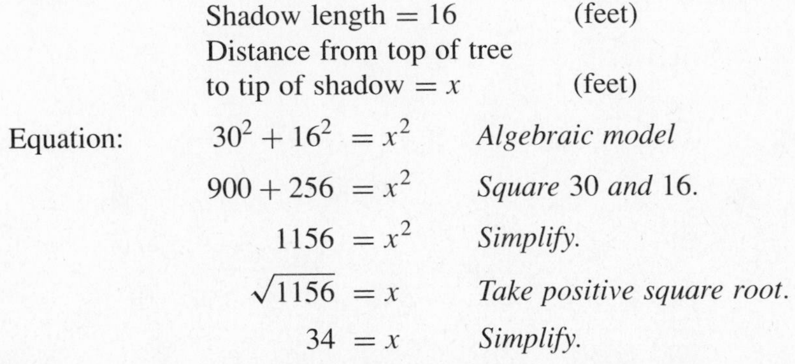

Labels:
Tree height = 30 (feet)
Shadow length = 16 (feet)
Distance from top of tree to tip of shadow = x (feet)

Equation:

$30^2 + 16^2 = x^2$	*Algebraic model*
$900 + 256 = x^2$	*Square* 30 *and* 16.
$1156 = x^2$	*Simplify.*
$\sqrt{1156} = x$	*Take positive square root.*
$34 = x$	*Simplify.*

The distance from the top of the tree to the tip of the shadow is 34 feet. Since x is the hypotenuse of the right triangle, the answer is reasonable.

Reteach
Chapter 9

Name ______________________

Guidelines:

- In a right triangle, the square of the length of the hypotenuse is equal to the sum of the squares of the lengths of the legs (Pythagorean Theorem).
- A set of three positive integers, a, b, and c, that satisfy the equation $a^2 + b^2 = c^2$ is a Pythagorean triple.

EXERCISES

In Exercises 1–4, use the Pythagorean Theorem to find the value of x.

1. 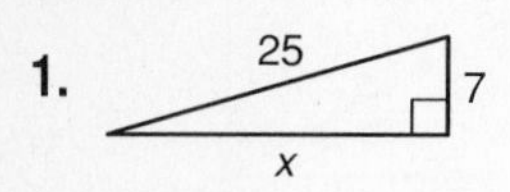

2. 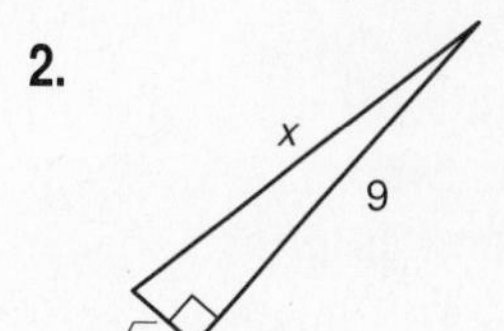

3. 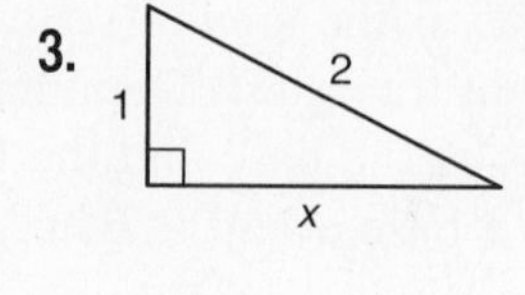

4. 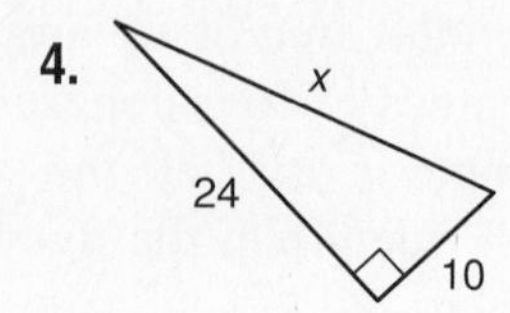

5. A weather balloon is seen directly above a point on the ground which is 350 feet from an observer, as shown at the right. If the balloon is 840 feet high, write and solve an equation for the distance from the balloon to the base of the observer.

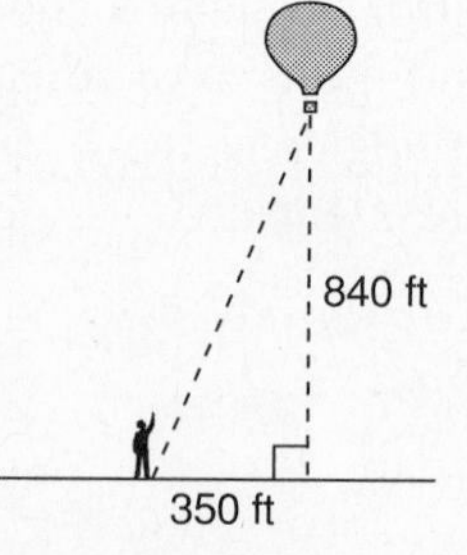

Verbal model: (Balloon height)^2 + (Distance from the ground point to the base of the observer)^2 = (Distance from the balloon to the base of the observer)^2

Labels: Balloon height = 840 (feet)

Distance from the ground point to the base of the observer = 350 (feet)

Distance from the balloon to the base of the observer = x (feet)

Reteach

Chapter 9

Name ______________________

What you should learn:

9.3	How to prove the converse of the Pythagorean Theorem and use the Pythagorean converse to solve problems

Correlation to Pupil's Textbook:

Mid-Chapter Self-Test (p. 447)
Exercises 10–13, 17–20

Chapter Test (p. 473)
Exercises 7–10

Examples — *Proving and Using the Pythagorean Converse*

a. In the figure at the right, $PQRS$ is a parallelogram and $PR = QS = 35$. Write a paragraph proof to prove that $PQRS$ is a rectangle.

Given: $\square PQRS$, $PS = QR = 21$, $PQ = RS = 28$,
$PR = 35$

Prove: $\square PQRS$ is a rectangle.

Proof:

- Square segment lengths $(PS)^2 = 21^2 = 441$, $(SR)^2 = 28^2 = 784$, and $(PR)^2 = 35^2 = 1225$.
- Since $441 + 784 = 1225$, you can substitute $(PS)^2 + (SR)^2 = (PR)^2$.
- By the Pythagorean Theorem Converse, $\triangle PSR$ is a right triangle.
- Similarly, you can show that $\triangle PQR$, $\triangle SPQ$, and $\triangle QRS$ are right triangles.
- By the definition of a right triangle, $\angle SPQ$, $\angle PQR$, $\angle QRS$, and $\angle PSR$ are right angles.
- Since the parallelogram $PQRS$ has four right angles, $PQRS$ is a rectangle.

b. You are examining a picture frame to decide if it is rectangular (if it has a right angle). You measure $\overline{XZ}$, $\overline{WX}$, and $\overline{WZ}$ and square their lengths. How do you know if the triangle $\triangle XWZ$ is acute, obtuse, or right?

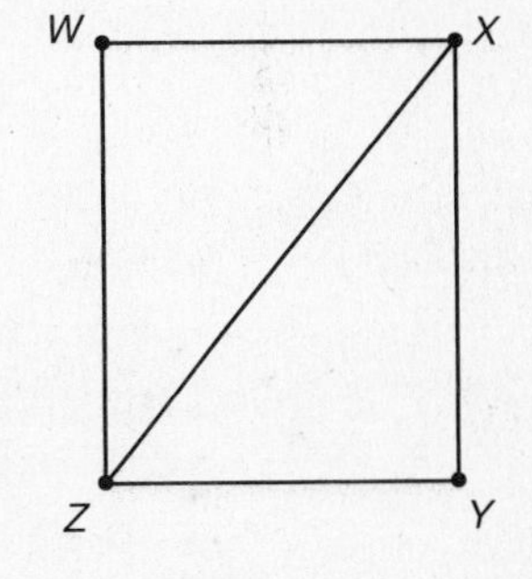

If $(XZ)^2 < (WX)^2 + (WZ)^2$, then $\triangle XWZ$ is acute.
If $(XZ)^2 > (WX)^2 + (WZ)^2$, then $\triangle XWZ$ is obtuse.
If $(XZ)^2 = (WX)^2 + (WZ)^2$, then $\triangle XWZ$ is a right triangle and the picture frame is rectangular.

Guidelines:

- $\triangle ABC$ is a right triangle if and only if $a^2 + b^2 = c^2$.

EXERCISES

In Exercises 1–3, match the side lengths with the correct description.

a. Acute triangle **b.** Obtuse triangle **c.** Right triangle

1. 15, 36, 40 **2.** 15, 36, 39 **3.** 15, 36, 38

Reteach
Chapter 9

Name ______________________________

What you should learn:

9.4	How to find the lengths of sides of special right triangles and solve problems using special right triangles

Correlation to Pupil's Textbook:

Chapter Test (p. 473)

Exercises 12–15, 18, 19

Examples *Finding and Using Side-Length Ratios of Special Right Triangles*

a. For each triangle, find the side lengths.

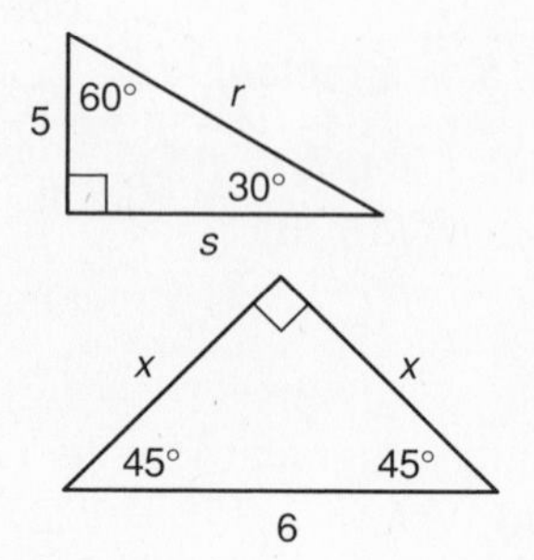

In a 30°-60°-90° triangle, the side-length ratios of short leg : long leg : hypotenuse are $1{:}\sqrt{3}{:}2$.

Since the longer leg is $\sqrt{3}$ times as long as the shorter leg, $s = 5\sqrt{3}$. Since the hypotenuse is twice as long as the shorter leg, $r = 10$.

In a 45°-45°-90° triangle, the side-length ratios of leg : leg : hypotenuse are $1{:}1{:}\sqrt{2}$. If the hypotenuse is $\sqrt{2}$ times as long as each leg, then divide by $\sqrt{2}$ to find the length of the leg. In this problem, then, $x = \dfrac{6}{\sqrt{2}}$ or $3\sqrt{2}$.

b. A 24-foot ladder leaning against a warehouse makes a 60° angle with the ground. How far up the side of the warehouse does the ladder reach?

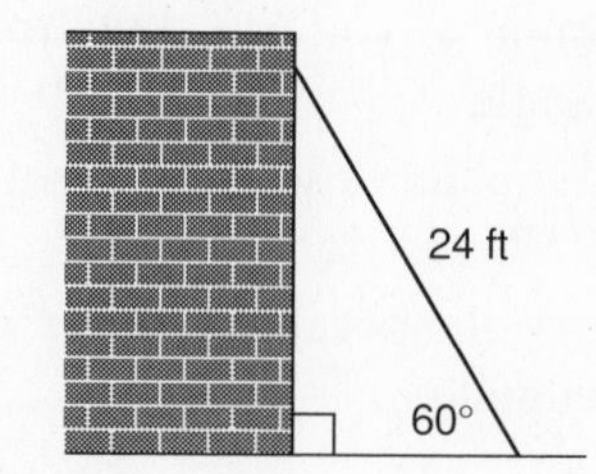

When the angle of elevation of the ladder is 60°, the distance up the side of the warehouse is the longer leg of a 30°-60°-90° triangle. If the hypotenuse is 24 feet, the shorter leg is 12 feet. Therefore, the distance up the side of the warehouse is $12\sqrt{3} \approx 21$ feet.

Guidelines:

- In a 45°-45°-90° triangle, the hypotenuse is $\sqrt{2}$ times as long as each leg.
- In a 30°-60°-90° triangle, the hypotenuse is twice as long as the shorter leg and the longer leg is $\sqrt{3}$ times as long as the shorter leg.

EXERCISES

In Exercises 1–4, find the value of each variable.

1.

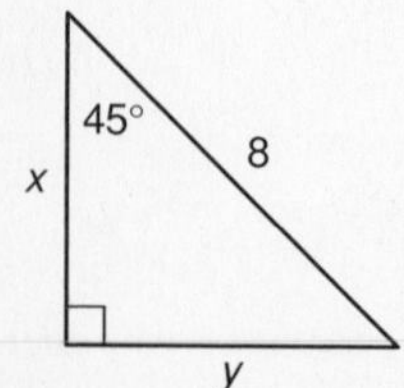

2.

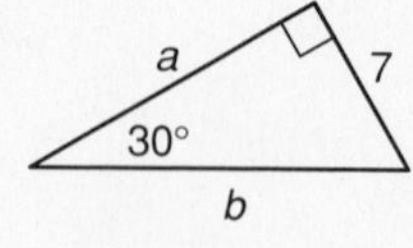

3.

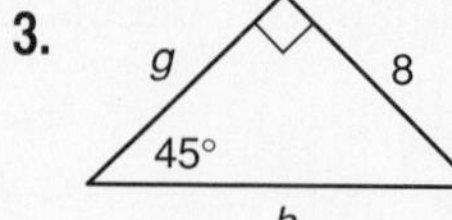

4.

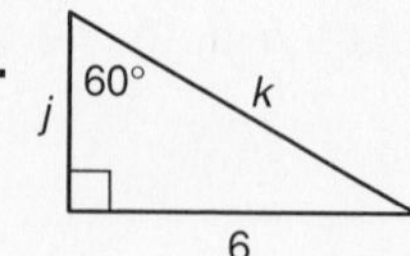

Reteach

Chapter 9

Name ______________________________

What you should learn:

9.5	How to find the sine, cosine, and tangent of an acute angle and use trigonometric ratios to solve real-life problems

Correlation to Pupil's Textbook:

Chapter Test (p. 473)

Exercises 20, 21

Examples *Finding Trigonometric Ratios*

a. For the triangle at the right, find the sine, cosine, and tangent of $\angle X$.

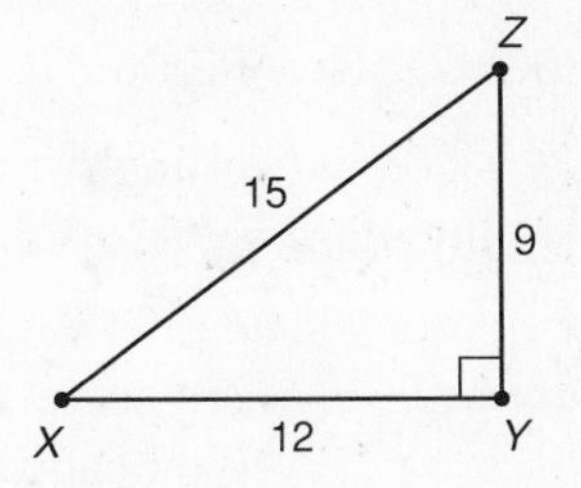

$$\sin X = \frac{\text{Side opposite } \angle X}{\text{Hypotenuse}} = \frac{9}{15} = \frac{3}{5}$$

$$\cos X = \frac{\text{Side adjacent to } \angle X}{\text{Hypotenuse}} = \frac{12}{15} = \frac{4}{5}$$

$$\tan X = \frac{\text{Side opposite } \angle X}{\text{Side adjacent to } \angle X} = \frac{9}{12} = \frac{3}{4}$$

b. Without using a calculator, find the sine, cosine, and tangent of $60°$.

Choose 1 as the length of the shorter leg. Since the side-length ratios are $1:\sqrt{3}:2$, the length of the longer leg is $\sqrt{3}$ and the length of the hypotenuse is 2.

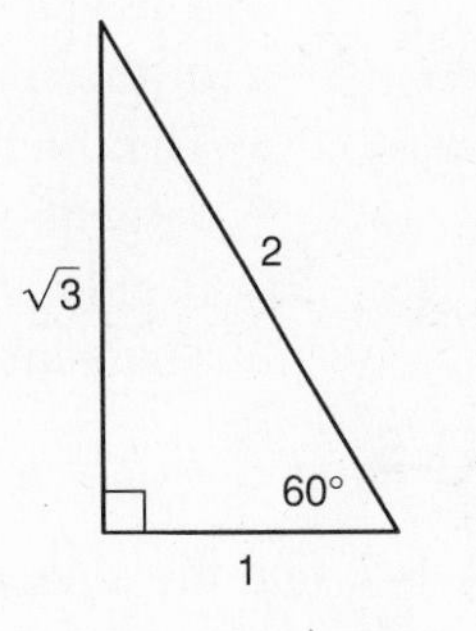

$$\sin 60° = \frac{\text{opp.}}{\text{hyp.}} = \frac{\sqrt{3}}{2}$$

$$\cos 60° = \frac{\text{adj.}}{\text{hyp.}} = \frac{1}{2}$$

$$\tan 60° = \frac{\text{opp.}}{\text{adj.}} = \frac{\sqrt{3}}{1}$$

c. A ramp designed for performing tricks on water skis has an angle of elevation of $20°$. If the length of the ramp is 11 feet, what is the length of the vertical drop (distance from the end of the ramp to the water)?

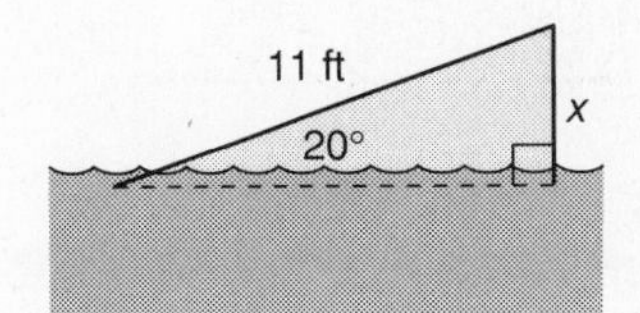

Use the sine ratio to find x, the length of the vertical drop.

$$\sin 20° = \frac{\text{opp.}}{\text{hyp.}}$$

$$0.342020 = \frac{x}{11}$$

$$3.76 \approx x$$

The length of the vertical drop is about 3.76 feet.

Reteach
Chapter 9

Name ______________________

d. A survey crew is planning a railroad line which rises to 1200 feet at a grade of 5°. Solve for x, the length of the railroad line.

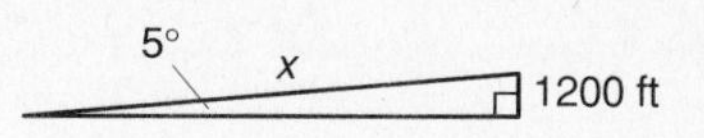

Use the sine ratio to find x, the length of the railroad line.

$$\sin 5^\circ = \frac{\text{opp.}}{\text{hyp.}}$$

$$0.087156 = \frac{1200}{x}$$

$$0.087156x = 1200$$

$$x \approx 13{,}768$$

The length of the railroad line is about 13,768 feet.

Guidelines:

- To solve a real-life problem involving a known angle of elevation, you can use a trigonometric ratio to find an unknown length.
- A trigonometric ratio is a ratio of the lengths of two sides of a right triangle.
- The value of a trigonometric ratio depends only on the measure of the acute angle of the triangle.
- The three basic trigonometric ratios are sine, cosine, and tangent, which are abbreviated as *sin*, *cos*, and *tan*, respectively.
- If only two sides of the right triangle are known, the Pythagorean Theorem should be used to determine the length of the third side.

EXERCISES

In Exercises 1–3, find the sine, cosine, and tangent of both acute angles of the right triangle.

1.

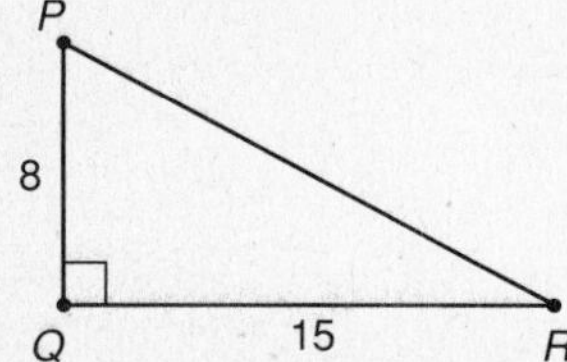

2.

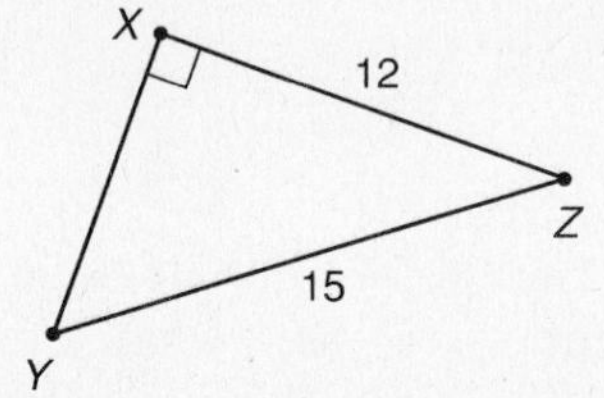

3.

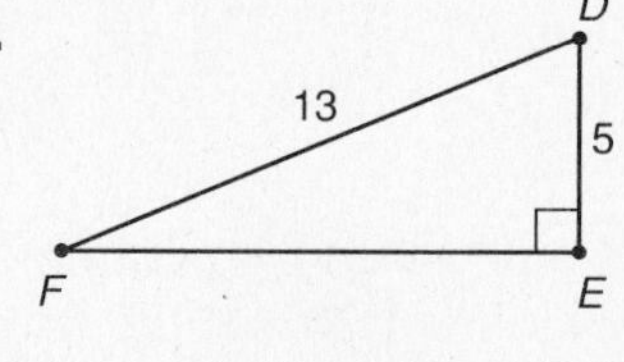

In Exercises 4–6, use a scientific calculator and/or trigonometric tables to find the lengths of the labeled sides.

4.

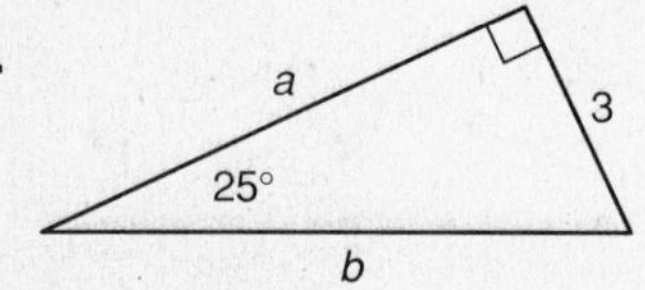

5.

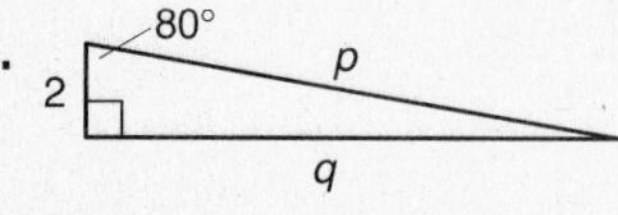

6.

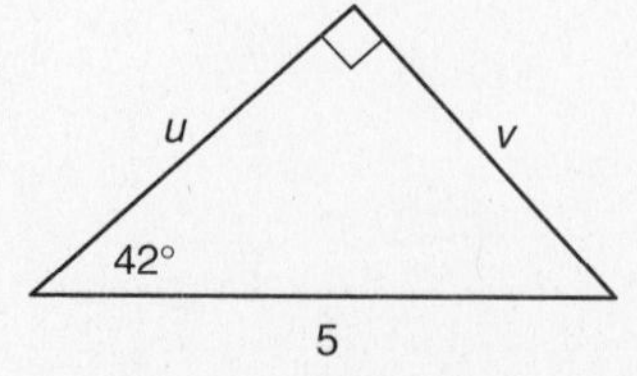

7. To measure the width, XY, of a river, a surveyor walks along the bank from point Y to point Z. She determines that $m\angle Z = 40^\circ$ and $ZY = 220$ feet. Explain how she can find XY, the width of the river.

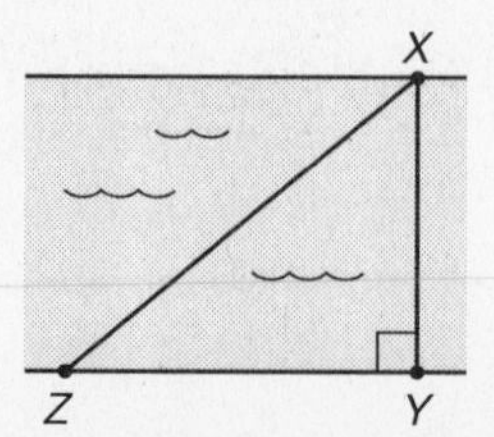

Reteach
Chapter 9

Name ______________________________

What you should learn:

9.6 How to solve a right triangle and use right triangles to solve real-life problems

Correlation to Pupil's Textbook:

Chapter Test (p. 473)
Exercises 1–6, 16, 17, 20, 21

Examples *Solving Right Triangles and Using Right Triangles in Real Life*

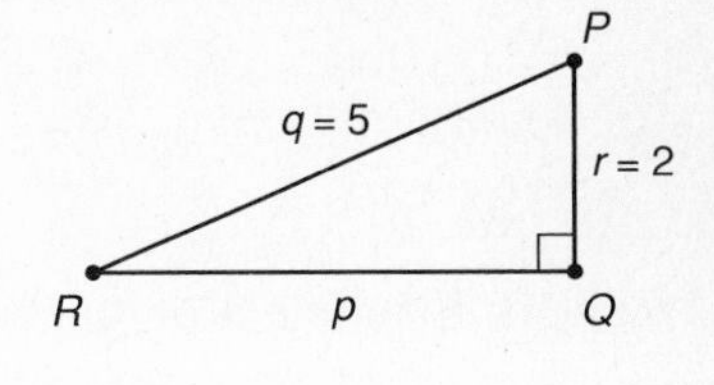

a. Solve the right triangle, shown at the right.

Use the Pythagorean Theorem to solve for p.

$p^2 + r^2 = q^2$ *Pythagorean Theorem*

$p^2 + 2^2 = 5^2$ *Substitute 2 for r and 5 for q.*

$p^2 + 4 = 25$ *Simplify.*

$p^2 = 21$ *Simplify.*

$p = \sqrt{21}$ *Take positive square root.*

Use a trigonometric ratio to find the measure of $\angle R$.

$$\sin R = \frac{\text{opp.}}{\text{hyp.}} = \frac{2}{5} = 0.4$$

Use your scientific calculator to obtain a decimal approximation of $m\angle R$. You determine that $m\angle R \approx 23.6°$. Since $\angle R$ and $\angle P$ are complementary, $m\angle P \approx 90° - 23.6° = 66.4°$.

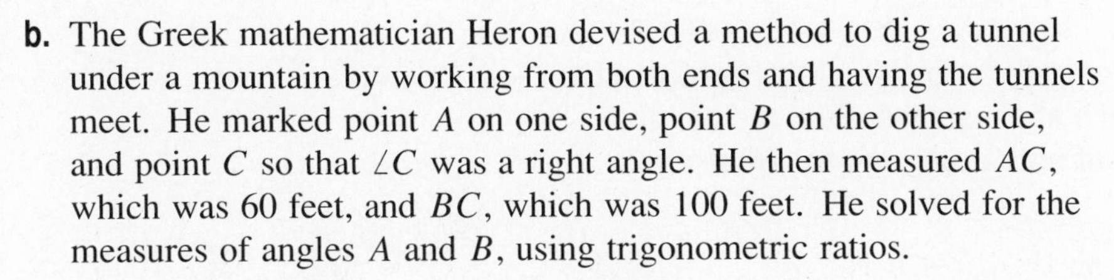

b. The Greek mathematician Heron devised a method to dig a tunnel under a mountain by working from both ends and having the tunnels meet. He marked point A on one side, point B on the other side, and point C so that $\angle C$ was a right angle. He then measured AC, which was 60 feet, and BC, which was 100 feet. He solved for the measures of angles A and B, using trigonometric ratios.

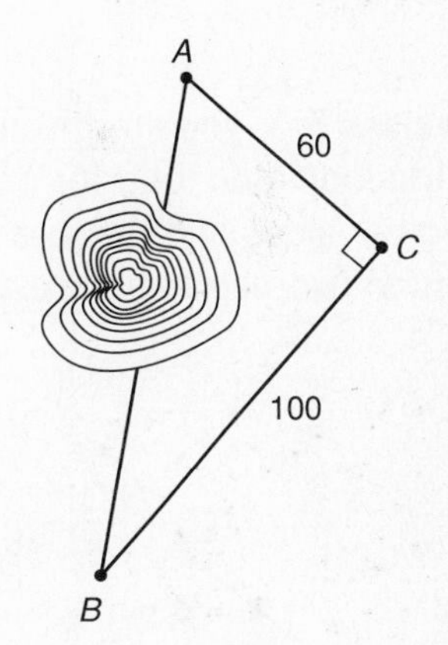

$$\tan B = \frac{\text{opp.}}{\text{adj.}} = \frac{60}{100} = 0.6 \qquad \tan A = \frac{\text{opp.}}{\text{adj.}} = \frac{100}{60} \approx 1.\overline{6}$$

$$m\angle B \approx 31° \qquad m\angle A \approx 59°$$

If the workers at point B followed a line 31° from $\overline{BC}$ and the workers at point A followed a line 59° from $\overline{CA}$, then the two tunnels met.

Reteach
Chapter 9

Name ______________________________

Guidelines:

- Solving a right triangle means to determine the measures of all six parts: three angles and three sides.
- You can solve a right triangle if you know
  1. the right angle and
  2. two side lengths, or one side length and one acute angle measure, or one side length and one trigonometric ratio.

EXERCISES

1. Rework Example a on page 89, using $r = 4$ and $q = 5$.

In Exercises 2–4, you are given the lengths of two sides of the right triangle. Find the length of the other side and the measures of both acute angles. Round your answers to two decimal places.

2.

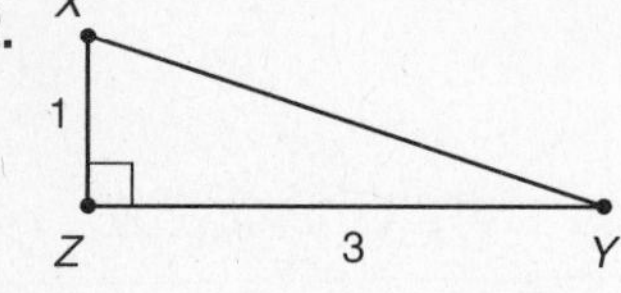

3.

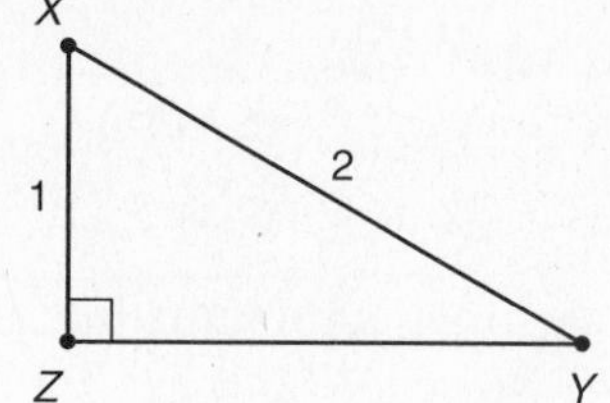

4.

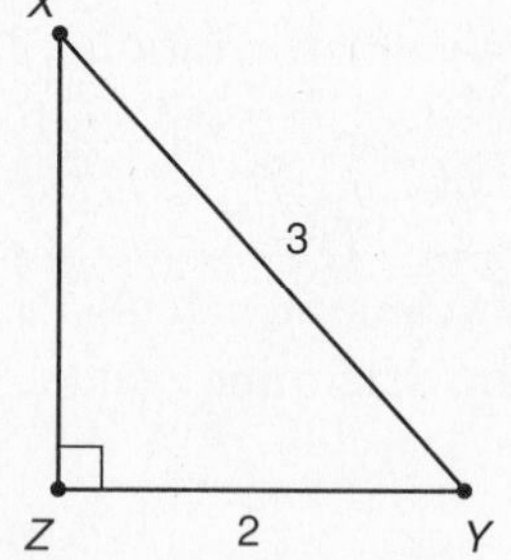

In Exercises 5–7, you are given the length of one side of the right triangle and a trigonometric ratio for one of the acute angles. Find the lengths of the other sides and the measures of both acute angles. Round your answers to two decimal places.

5. $\sin P = 0.781$

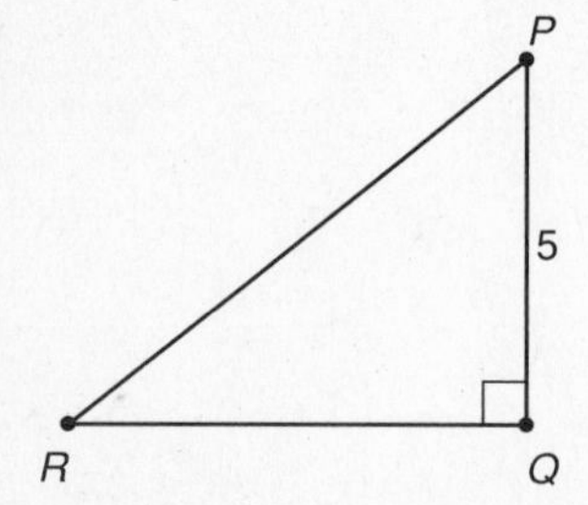

6. $\cos P = \frac{1}{3}$

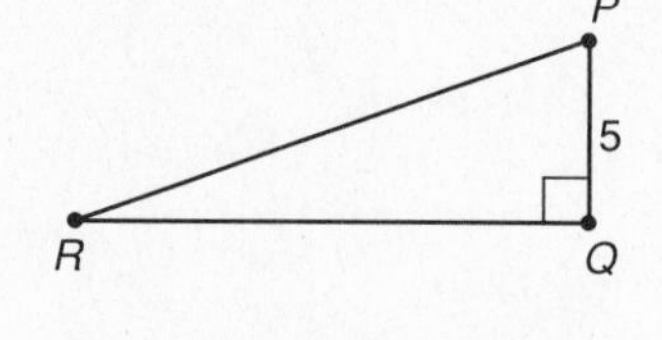

7. $\tan P = 1$

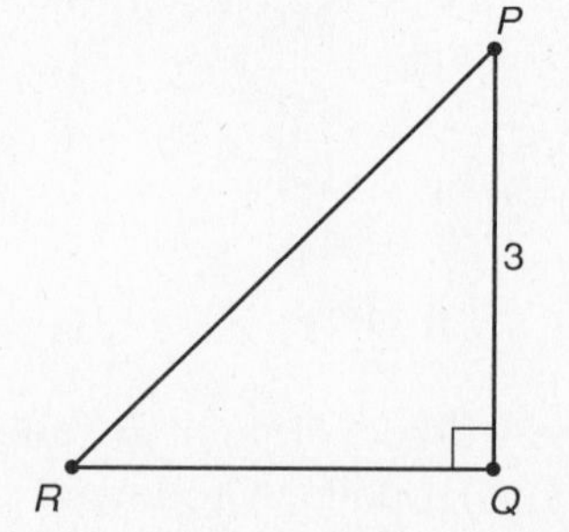

Reteach

Chapter 10

Name ______________________________

What you should learn:

10.1 How to use vocabulary of circles and describe circles in real-life situations

Correlation to Pupil's Textbook:

Mid-Chapter Self-Test (p. 498) Exercises 1–10

Chapter Test (p. 529) Exercises 3, 5–8

Examples *Using the Vocabulary of Circles and Describing Circles in Real Life*

a. Using the figure at the right, identify each of the following.

Answers:

1. radius (segment that has the center as one endpoint and a point on the circle as the other endpoint) — $\overline{RQ}$ and $\overline{SQ}$
2. center (a given point from which every point is equidistant) — Q
3. chord (segment whose endpoints are on the circle) — $\overline{RS}$ and $\overline{UV}$
4. diameter (chord that passes through the center) — $\overline{RS}$
5. interior point (a point inside the circle) — P and Q
6. exterior point (a point outside the circle) — T

b. Using the figure at the right, identify each of the following.

Answers:

1. tangent (a line that intersects a circle at exactly one point) — $\overleftrightarrow{CD}, \overleftrightarrow{EF}$
2. point of tangency (point of intersection of a tangent line and a circle) — C, D, E, F
3. secant (line that intersects a circle at two points) — $\overleftrightarrow{AB}$
4. common tangent (a line tangent to two circles) — $\overleftrightarrow{CD}, \overleftrightarrow{EF}$

c. The Greeks were interested in finding the distance from the earth to the moon using trigonometry. In the figure at the right, QM is the distance from the surface of the earth to the surface of the moon. Identify each of the following for $\odot E$.

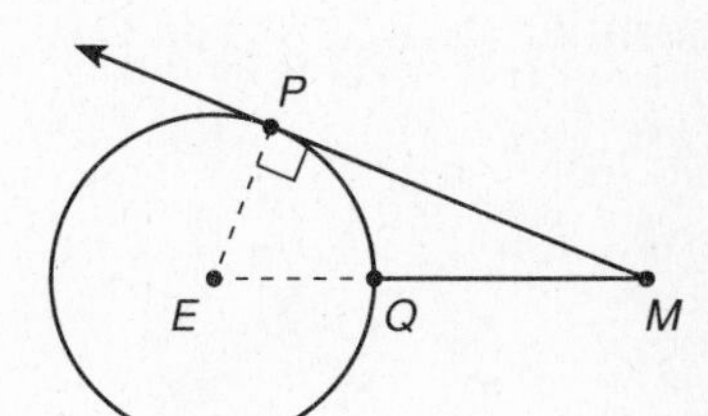

1. radius *Answer:* $\overline{EQ}, \overline{EP}$
2. center *Answer:* E
3. exterior point *Answer:* M
4. tangent line *Answer:* $\overleftrightarrow{PM}$
5. point of tangency *Answer:* P

d. The figure at the right illustrates a lunar eclipse, in which $\odot S$ represents the sun and $\odot E$ represents the earth. A part of the path of the moon, M, around the earth is shown by the dotted curve. Identify each of the following for $\odot S$ and $\odot E$.

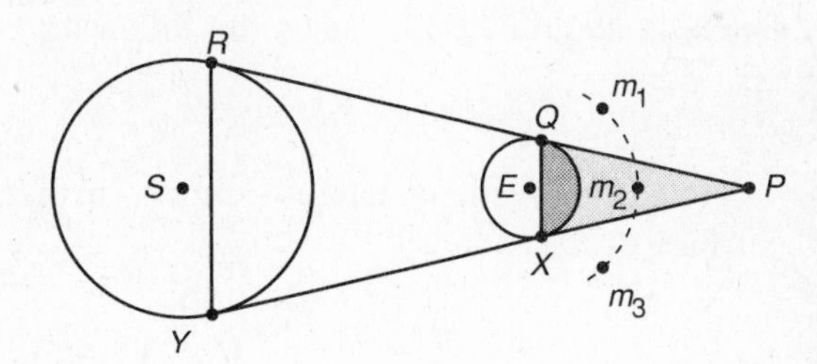

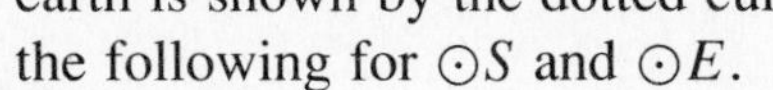

1. common tangent *Answer:* $\overleftrightarrow{RQ}, \overleftrightarrow{XY}$
2. chord *Answer:* $\overline{RY}, \overline{QX}$

Reteach
Chapter 10

Name ______________________________

Guidelines:

- Two circles can
  - — be internally or externally *tangent* (one point of intersection).
  - — intersect in two points.
  - — have the same center (*concentric*).
  - — be congruent (congruent radii or congruent diameters).
- You can use circles and tangent lines to solve real-life problems such as finding gear ratios and estimating distances in astronomy.

EXERCISES

1. Use a compass to draw two circles that are

a. internally tangent. b. externally tangent. c. intersecting.

d. concentric. e. congruent.

2. Solve each of the following using the fact that the diameter, d, of a circle is twice its radius, r.

a. Given $r = 5.4$ cm, find d. b. Given $d = 7.8$ in., find r.

c. Given $d = \frac{5}{3}$ in., find r. d. Given $r = \sqrt{2}$ cm, find d.

In Exercises 3–12, use the figure at the right to state the geometric term that best describes the given notation.

3. $\overline{QR}$ **4.** $\overleftrightarrow{WX}$

5. $\overleftrightarrow{ST}$ **6.** P

7. V **8.** $\overline{OU}$

9. $\overline{WX}$ **10.** $\overleftrightarrow{UQ}$

11. S **12.** Z

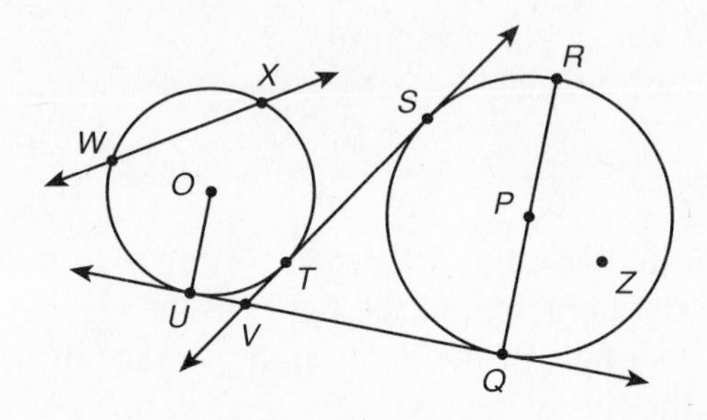

Use the figure in Example c on page 91, where $\angle EPM$ is a right angle. If you know that the radius of the earth, $\overline{EP}$, is about 4000 miles and $m\angle PEM$ is about 89.067°, solve the following.

13. Solve right triangle $\triangle EPM$ for EM.

14. Solve for QM, the distance from the surface of the earth at point Q to the surface of the moon.

Reteach
Chapter 10

Name ______________________________

What you should learn:

10.2 How to use properties of tangents to solve problems in geometry and in real life

Correlation to Pupil's Textbook:

Mid-Chapter Self-Test (p. 498) Exercises 11–16, 23

Chapter Test (p. 529) Exercise 1

Examples *Using Properties of Tangents to Solve Problems*

a. Use the figure at the right in which $\overleftrightarrow{BA}$ and $\overleftrightarrow{BC}$ are tangents to $\odot P$ at radii $\overline{PA}$ and $\overline{PC}$. Find the lengths of $\overline{AB}$ and $\overline{BC}$.

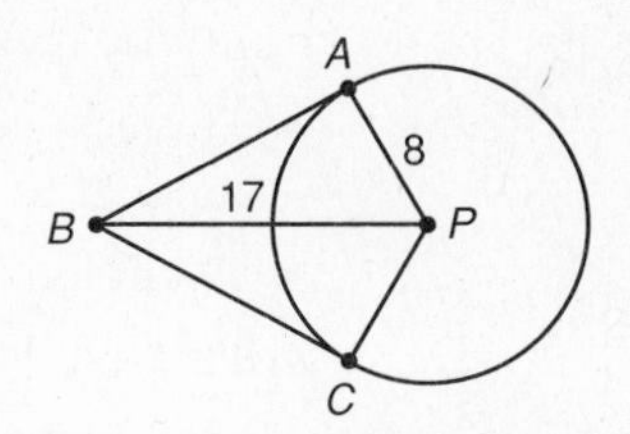

Using Theorem 10.1, $\overleftrightarrow{AB} \perp \overline{PA}$ and $\overleftrightarrow{BC} \perp \overline{PC}$. In follows that $\angle BAP$ and $\angle BCP$ are right angles. In right triangles $\triangle ABP$ and $\triangle CBP$, use Pythagorean triples to find $AB = BC = 15$.

b. Use the figure at the right in which $\overline{CA}$ is tangent to $\odot M$ at radius $\overline{MA}$ and $m\angle C$ is 0.25°. Solve for MA, the length of the radius of the moon.

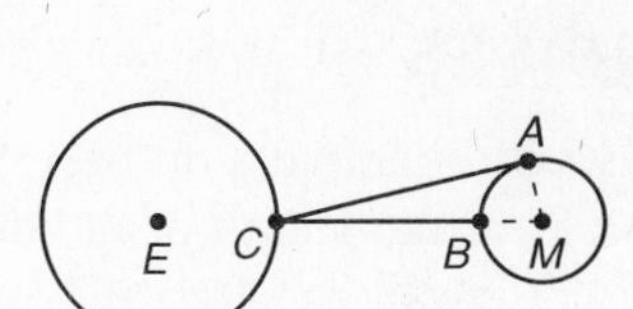

Figure not drawn to scale

You solved for CB, the distance between the surfaces of the earth and moon in Reteach 10.1 Exercise 14. You determined that $CB \approx 241{,}652$ miles. Since $\overline{CA}$ is tangent to $\odot M$ at radius $\overline{MA}$, $\overline{CA} \perp \overline{MA}$ and it follows that $\angle A$ is a right angle. Also, $CM = CB + MB$. Since $MB = MA$, $CM = CB + MA$. Solving right triangle $\triangle ACM$ for MA,

$$\sin C = \frac{MA}{CM}$$

$$\sin 0.25^\circ = \frac{MA}{241{,}652 + MA}$$

$$0.0043633 = \frac{MA}{241{,}652 + MA}$$

$$1054.402 + 0.0043633MA = MA$$

$$1054.402 = 0.9956367MA$$

$$\frac{1054.402}{0.9956367} = MA$$

$$1059 \approx MA.$$

The length of the radius of the moon is about 1059 miles.

c. In the figure at the right, each side of regular hexagon $ABCDEF$ is tangent to $\odot X$. Also, each vertex of regular hexagon $PQRSTU$ is on $\odot X$. Match the following statements.

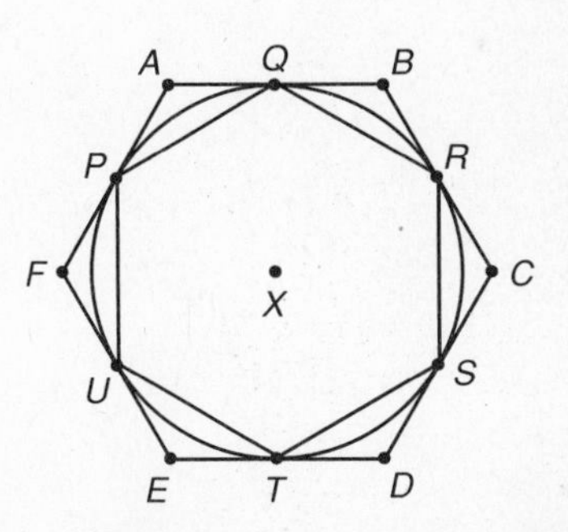

1. Circle $\odot X$ is [?] polygon $ABCDEF$.
2. Circle $\odot X$ is [?] polygon $PQRSTU$.

a. circumscribed about
b. inscribed in

Answers: 1. b 2. a

Reteach

Chapter 10

Name ______________________________

Guidelines:

- If a line is tangent to a circle, then it is perpendicular to the radius drawn to the point of tangency.
- In a plane, if a line is perpendicular to a radius of a circle at its endpoint on the circle, then the line is tangent to the circle.
- If two segments from the same exterior point are tangent to a circle, then they are congruent.
- A circle is inscribed in a polygon if each side of the polygon is tangent to the circle.
- A circle is circumscribed about a polygon if each vertex of the polygon is on the circle.

EXERCISES

1. Rework Example a on page 93 with different measurements: $AB = 17$ and $BP = \sqrt{353}$. Solve for the length of radius $\overline{PC}$.

2. Use the diagram below to find PD.

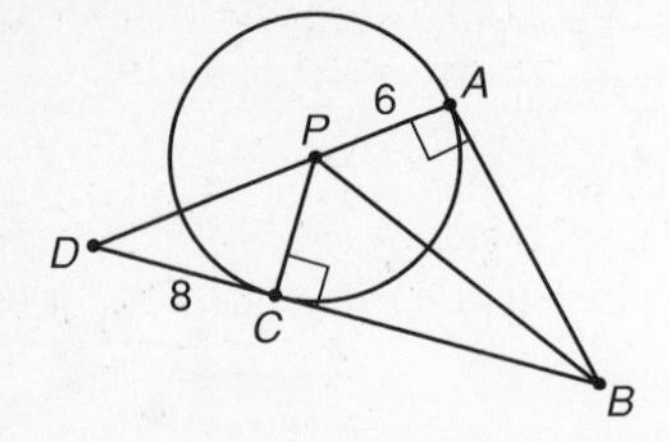

3. Use the diagram below to find BC.

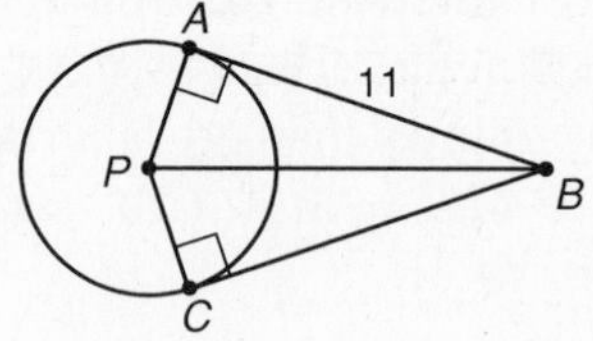

4. Use the diagram below to find $m\angle PQR$. Line ℓ is tangent to circle P at point Q.

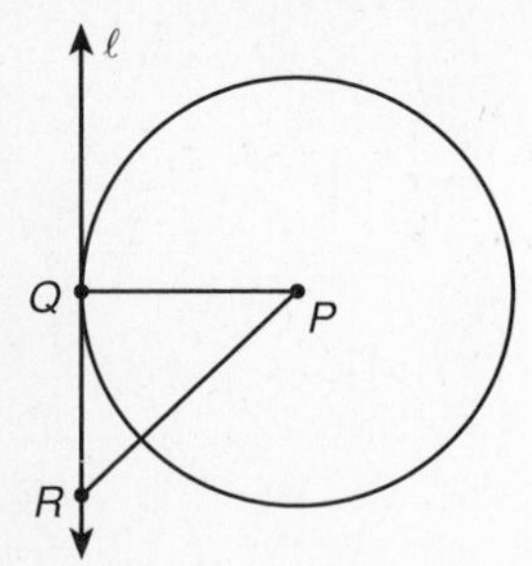

5. In the diagram below, the radius of circle P is 7. Explain why $\overleftrightarrow{QR}$ is tangent to circle P.

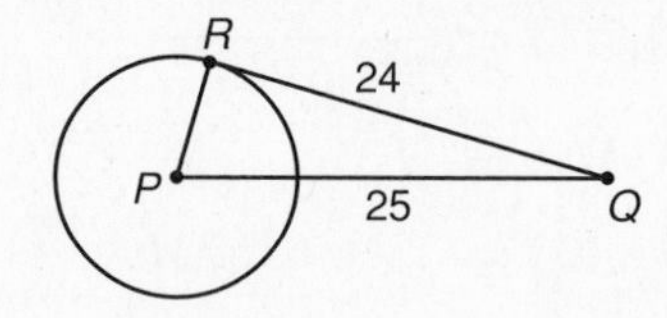

Reteach

Chapter 10

Name ______________________________

What you should learn:

10.3 How to measure central angles and arcs of circles and use the measures to solve real-life problems

Correlation to Pupil's Textbook:

Mid-Chapter Self-Test (p. 498) Exercises 17–22, 24

Chapter Test (p. 529) Exercise 13

Examples *Measuring Central Angles and Arcs of Circles and Using Arcs*

a. Match each of the following, using the figure at the right.

1. central angle (vertex is the center of the circle) — a. $\widehat{CQD}$
2. minor arc (formed by a central angle with measure less than 180°) — b. 130°
3. major arc (formed by all points of circle that lie in the exterior of a central angle) — c. $\angle CPD$
4. measure of minor arc (measure of central angle) — d. 230°
5. measure of major arc (difference of 360° and measure of the associated minor arc) — e. $\widehat{CD}$

Answers: 1. c 2. e 3. a 4. b 5. d

b. 1. Use the figures at the right to find $m\widehat{AC}$ and $m\widehat{ADC}$.

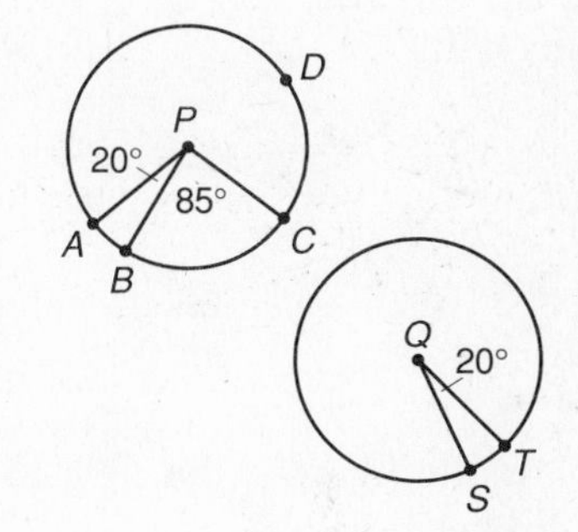

By the Arc Addition Postulate, $m\widehat{AC} = m\widehat{AB} + m\widehat{BC}$. Since $\widehat{AB}$ and $\widehat{BC}$ are minor arcs, their measures are the measures of their central angles, which are 20° and 85°. Therefore, $m\widehat{AC} = 20° + 85° = 105°$. The measure of major arc $\widehat{ADC}$ is $360° - m\widehat{AC} = 255°$.

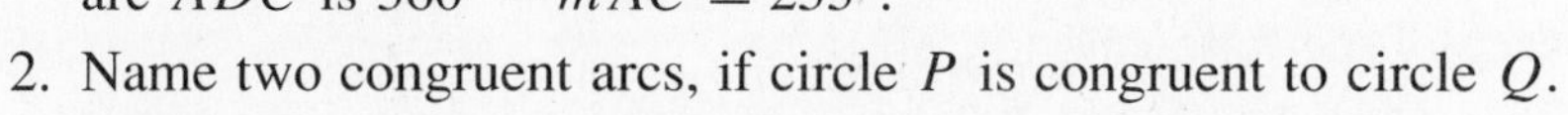

2. Name two congruent arcs, if circle P is congruent to circle Q.

In the same circle, or in congruent circles, two arcs are congruent if and only if their central angles are congruent. Since $m\angle APB = m\angle SQT = 20°$, $\angle APB \cong \angle SQT$. Circles $\odot P$ and $\odot Q$ are congruent, therefore, $\widehat{AB} \cong \widehat{ST}$.

c. Construct a circle graph for Americans' favorite bicycle colors, using the following percentages:

black-24%, blue-23%, red-22%, white-8%, silver-5%, other-18%

Multiply each percentage by 360° to determine the measure of each central angle.

black	$0.24 \times 360° \approx 86°$	white	$0.08 \times 360° \approx 29°$
blue	$0.23 \times 360° \approx 83°$	silver	$0.05 \times 360° \approx 18°$
red	$0.22 \times 360° \approx 79°$	other	$0.18 \times 360° \approx 65°$

Use a compass, protractor, and straightedge to draw the graph.

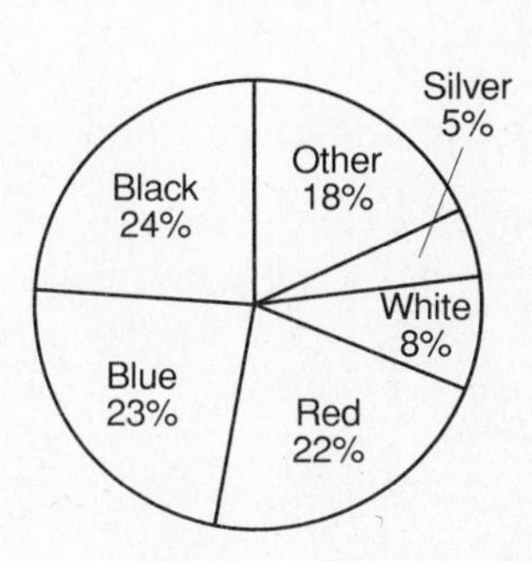

Reteach
Chapter 10

Name ______________________________

Guidelines:

- Minor arcs are denoted by two letters, as in $\overset{\frown}{BC}$.
- The measure of a minor arc is defined to be the measure of its central angle.
- Major arcs are denoted by three letters, as in $\overset{\frown}{BDC}$.
- The measure of a major arc is defined to be the difference between 360° and the measure of its associated minor arc.
- The measure of an arc formed by two adjacent arcs is the sum of the measures of the two arcs. (Arc Addition Postulate)
- In the same circle, or in congruent circles, two arcs are congruent if and only if their central angles are congruent.

EXERCISES

In Exercises 1–6, $\overline{QS}$ is a diameter. Find the indicated measure.

1. $m\overset{\frown}{ST}$ **2.** $m\overset{\frown}{RS}$ **3.** $m\overset{\frown}{RSQ}$

4. $m\angle QPT$ **5.** $m\overset{\frown}{RT}$ **6.** $m\overset{\frown}{RST}$

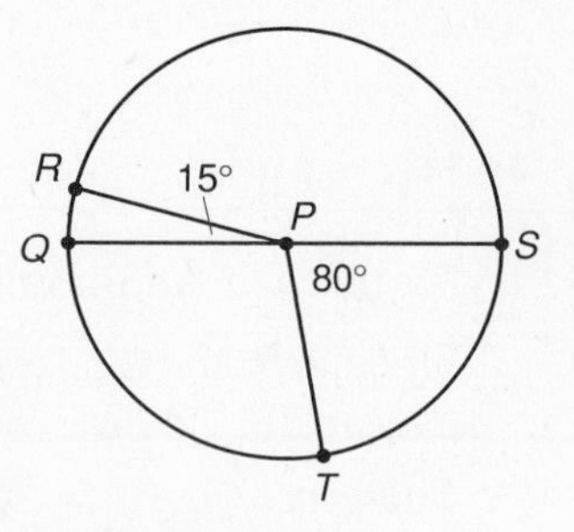

7. Use the figure at the right to name each of the following for circle P.

a. semicircle

b. a central angle with measure 100°

c. $m\overset{\frown}{WY}$

d. a minor arc

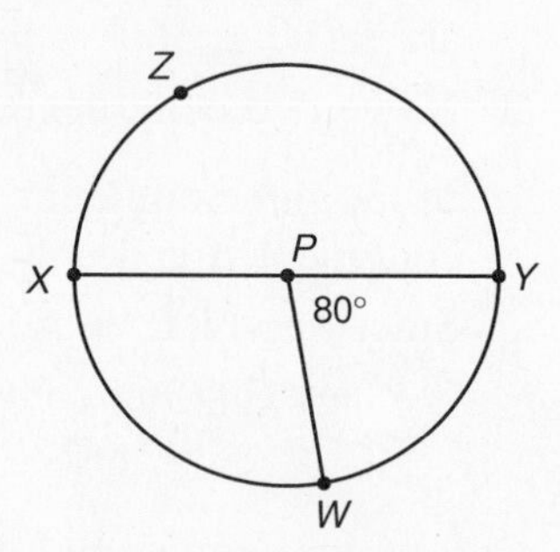

8. Use the figure at the right to name each of the following for circle Q.

a. two congruent minor arcs

b. $m\overset{\frown}{AC}$

c. a major arc with measure 220°

d. $m\overset{\frown}{AD}$

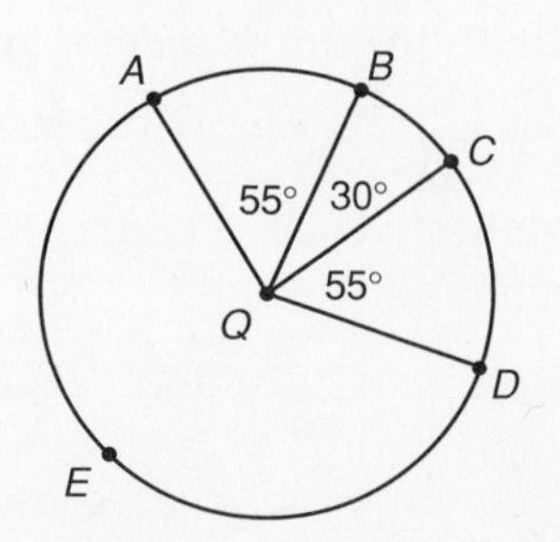

Reteach

Chapter 10

Name ______________________________

What you should learn:

10.4 How to use properties of arcs and chords to solve problems in geometry and in real life

Correlation to Pupil's Textbook:

Chapter Test (p. 529)
Exercises 2, 4, 9, 11, 12, 19

Examples — *Using Properties of Arcs and Chords to Solve Problems*

a. In the figure at the right, diameter $\overline{AC}$ is perpendicular to $\overline{BE}$, $\widehat{AG} \cong \widehat{FE}$, $PQ = PR$. Complete each congruence statement and justify your answer by stating the appropriate theorem.

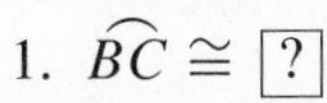

1. $\widehat{BC} \cong$ [?]

   Answer: $\widehat{CE}$ If diameter $\overline{AC}$ is perpendicular to chord $\overline{BE}$, then diameter $\overline{AC}$ bisects the chord $\overline{BE}$ and its arc $\widehat{BE}$.

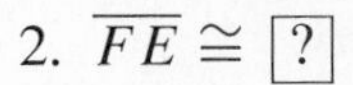

2. $\overline{FE} \cong$ [?]

   Answer: $\overline{AG}$ In $\odot P$, if two minor arcs $\widehat{AG}$ and $\widehat{FE}$ are congruent, then corresponding chords $\overline{FE}$ and $\overline{AG}$ are congruent.

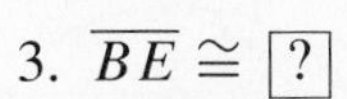

3. $\overline{BE} \cong$ [?]

   Answer: $\overline{FD}$ In $\odot P$, if two chords $\overline{BE}$ and $\overline{FD}$ are equidistant from the center P, then chords $\overline{BE}$ and $\overline{FD}$ are congruent.

b. Use the figure at the right to complete each congruence statement.

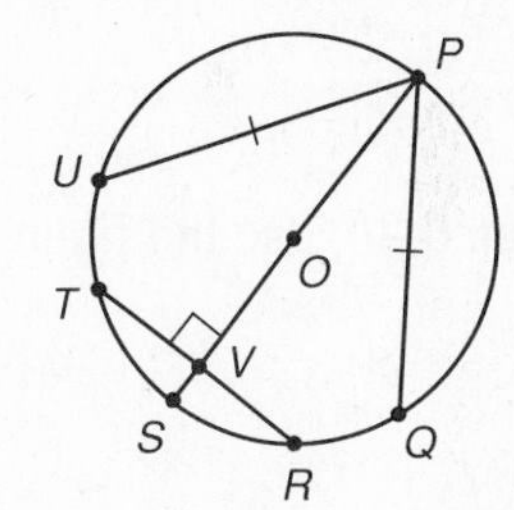

1. [?] $\cong \overline{VR}$ *Answer:* $\overline{TV}$
2. $\widehat{UP} \cong$ [?] *Answer:* $\widehat{QP}$
3. [?] $\cong \widehat{SR}$ *Answer:* $\widehat{TS}$

c. NASA launched a laser geodynamic satellite named Lageos in 1976. Using laser beam techniques, very precise measurements have improved NASA's forecasting of earthquakes. Suppose that Lageos located points X, Y, and Z where seismographs recorded equal earthquake readings. Explain how to locate point E, the epicenter.

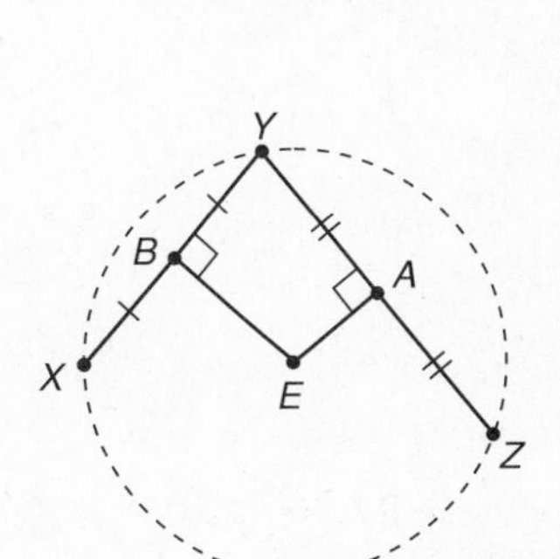

Construct $\overline{EA}$, the perpendicular bisector of $\overline{YZ}$, and $\overline{EB}$, the perpendicular bisector of $\overline{XY}$. The center of the circle through X, Y, and Z is E, the point of intersection of $\overline{EB}$ and $\overline{EA}$. Because radii $\overline{EX}$, $\overline{EY}$, and $\overline{EZ}$ are congruent, epicenter E is equidistant from locations X, Y, and Z.

Reteach
Chapter 10

Name ____________________

Guidelines:

- By constructing perpendicular bisectors, you can locate the center of the circumscribed circle that contains three given points.
- If chord $\overline{AB}$ is a perpendicular bisector of another chord, then $\overline{AB}$ is a diameter.
- In a circle or in congruent circles, two minor arcs are congruent if and only if their corresponding chords are congruent.
- If a diameter of a circle is perpendicular to a chord, then the diameter bisects the chord.
- In a circle or in congruent circles, two chords are congruent if and only if they are equidistant from the center.

EXERCISES

In the figure at the right, $\overline{XY} \perp \overline{ST}$, $PR = PW$. Complete each congruence statement.

1. $\overline{SR} \cong$ [?] **2.** $\widehat{XT} \cong$ [?] **3.** [?] $\cong \overline{UV}$

4. $\widehat{ST} \cong$ [?] **5.** [?] $\cong \overline{XT}$

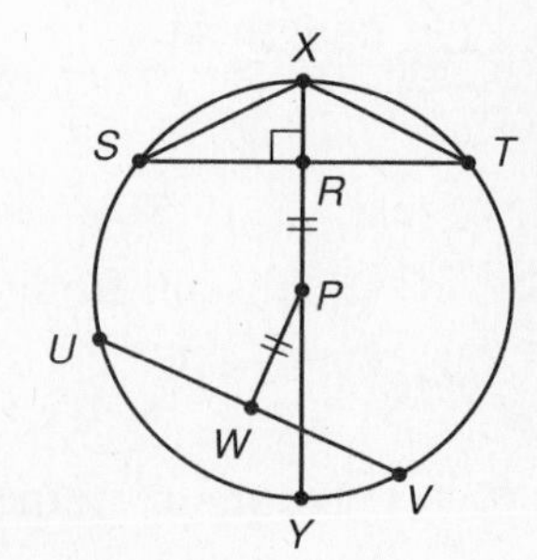

In Exercises 6–10, use the figure at the right to match equal measures.

6. CE a. 4

7. $m\widehat{ED}$ b. 8

8. CF c. 5

9. $m\widehat{AB}$ d. $m\widehat{DC}$

10. DC e. $m\widehat{EC}$

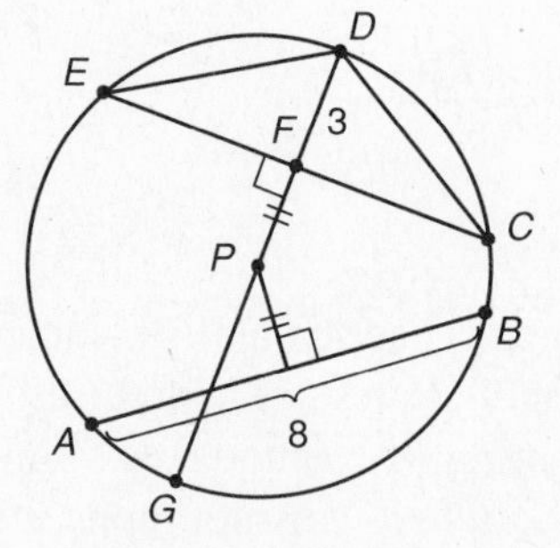

Reteach
Chapter 10

Name ______________________________

What you should learn:

10.5	How to use inscribed angles to solve problems in geometry and in real life

Correlation to Pupil's Textbook:

Chapter Test (p. 529)
Exercises 10, 13–15

Examples *Using Inscribed Angles to Solve Problems*

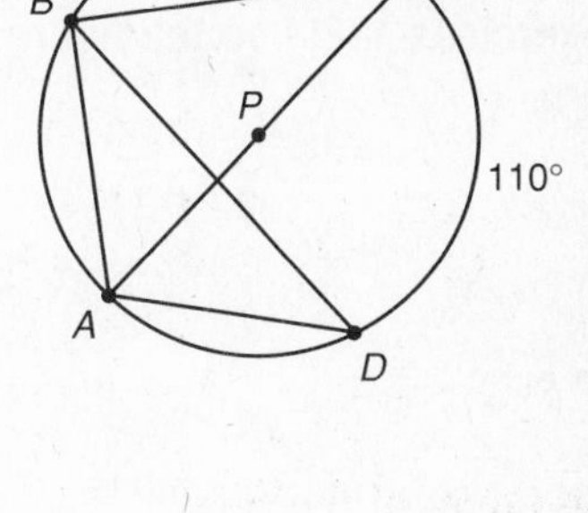

a. In the figure at the right, $m\overset{\frown}{CD} = 110°$. Complete each statement and justify your answer by stating the appropriate theorem or definition.

1. $\angle CAD$ is an [?] angle of $\odot P$.

   Answer: inscribed An angle $\angle CAD$ is an inscribed angle of a circle if $\overline{AC}$ and $\overline{AD}$ are chords of the circle.

2. $\overset{\frown}{CD}$ is called the [?] arc of inscribed $\angle CAD$.

   Answer: intercepted The arc that lies in the interior of inscribed $\angle CAD$ is called the intercepted arc of the angle.

3. $\angle$ [?] is a right angle.

   Answer: ABC Angle $\angle ABC$ that is inscribed in a circle is a right angle if and only if its corresponding arc, $\overset{\frown}{ADC}$, is a semicircle.

4. $m\angle CBD =$ [?]

   Answer: 55° If $\angle CBD$ is inscribed in a circle, then its measure is half the measure of its intercepted arc, $\overset{\frown}{CD}$.

5. $\angle CBD \cong \angle$ [?]

   Answer: CAD If two inscribed angles $\angle CBD$ and $\angle CAD$ intercept the same arc $\overset{\frown}{CD}$, then the angles are congruent.

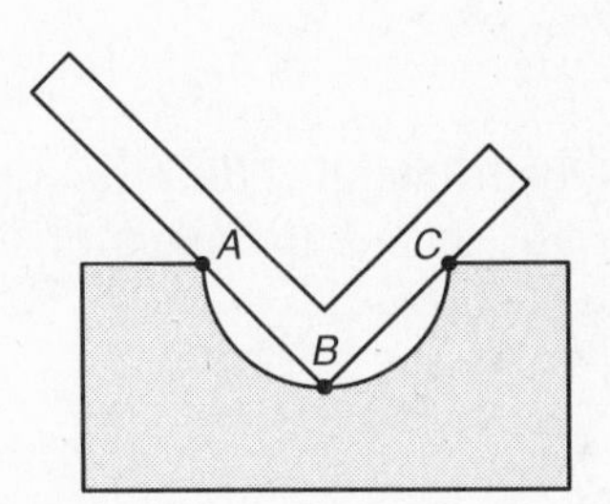

b. Suppose you are asked to test that a wooden beam has a groove that is a semicircle. How could you use a carpenter's square to test the groove?

Place the carpenter's square in the groove, as shown at the right. Move it around in all positions keeping the square directly across the groove. If the groove is a semicircle, the square will always touch at three points.

Reteach
Chapter 10

Name ________________________________

Guidelines:

- If an angle is inscribed in a circle, then its measure is half the measure of its intercepted arc.
- A quadrilateral can be inscribed in a circle if and only if its opposite angles are supplementary.

EXERCISES

In Exercises 1–8, use the figure at the right to find the measure of the angle or arc.

1. $\angle SPR$ **2.** $\widehat{PS}$ **3.** $\widehat{QR}$ **4.** $\angle QSR$

5. $\angle PSR$ **6.** $\widehat{PQR}$ **7.** $\angle SQR$ **8.** $\widehat{PQ}$

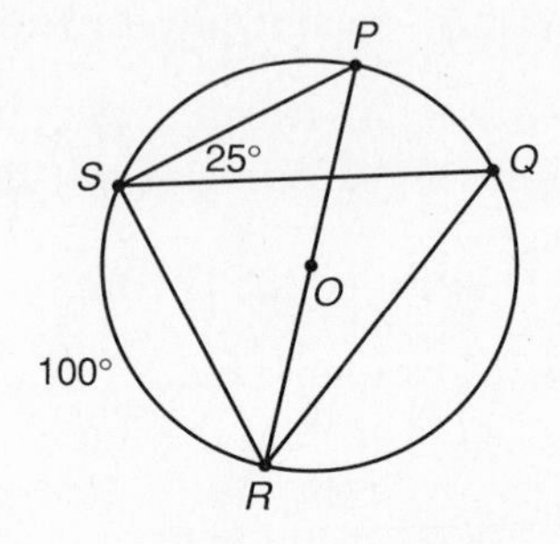

In Exercises 9–14, use the figure at the right to find the measure of the angle or arc. In the figure, $\widehat{SP} = 140°$ and $\widehat{RQ} = 50°$.

9. $m\angle PQR$ **10.** $m\widehat{SR}$

11. $m\angle SQP$ **12.** $m\angle RQS$

13. $m\widehat{PQ}$ **14.** $m\angle QSP$

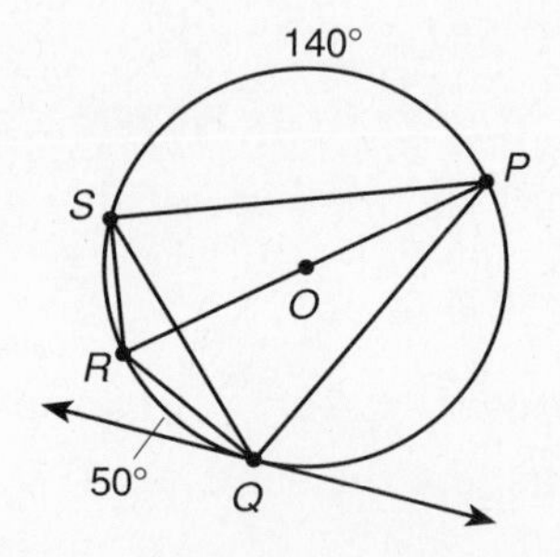

15. Quadrilateral $ABCD$ is inscribed in circle, P, as shown at the right. Name two pairs of supplementary angles.

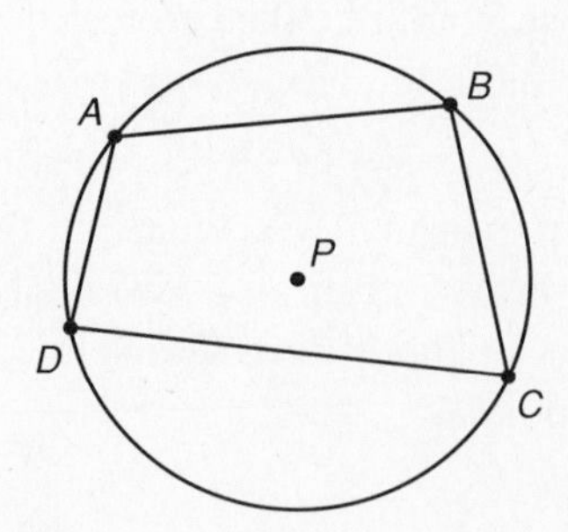

16. Can you inscribe all parallelograms in a circle? Explain your reasoning.

Reteach
Chapter 10

Name ______________________________

What you should learn:

10.6 How to measure angles formed by tangents, chords, and secants and use angle measures to solve real-life problems

Correlation to Pupil's Textbook:

Chapter Test (p. 529)
Exercises 14, 16

Examples *Findind Angle Measures and Using Angle Measures in Real Life*

a. In the figure at the right, find the measure of $\angle EFI$, $\angle EFJ$, $\angle CKD$, and $\angle CKH$.

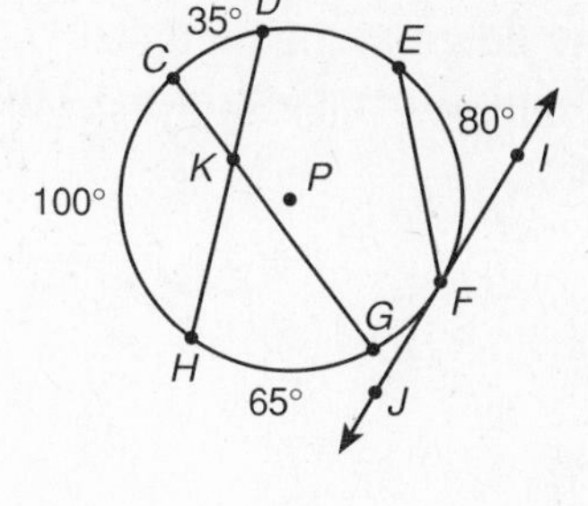

If a tangent $\overleftrightarrow{IJ}$ and chord $\overline{EF}$ intersect at point F on $\odot P$, then the measure of each angle formed is half the measure of its intercepted arc. Therefore,

$$m\angle EFI = \tfrac{1}{2}(m\overset{\frown}{EF}) = 40^\circ$$

$$\text{and } m\angle EFJ = \tfrac{1}{2}(m\overset{\frown}{EDF}) = 140^\circ.$$

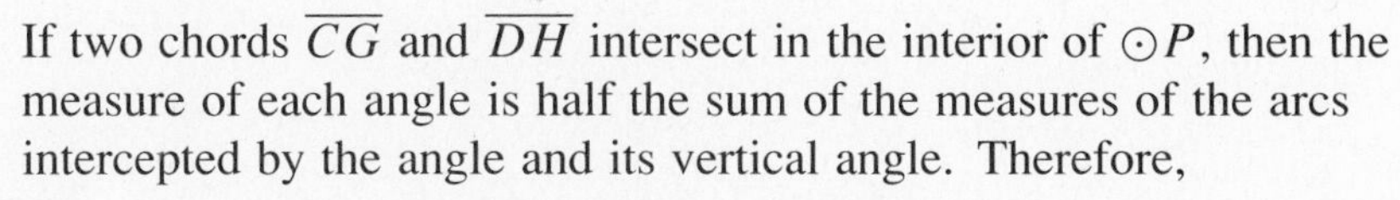

If two chords $\overline{CG}$ and $\overline{DH}$ intersect in the interior of $\odot P$, then the measure of each angle is half the sum of the measures of the arcs intercepted by the angle and its vertical angle. Therefore,

$$m\angle CKD = \tfrac{1}{2}(m\overset{\frown}{CD} + m\overset{\frown}{HG}) = 50^\circ$$

$$\text{and } m\angle CKH = \tfrac{1}{2}(m\overset{\frown}{CH} + m\overset{\frown}{DEG}) = 130^\circ.$$

b. In the figure at the right, express the measures of $\angle BAC$, $\angle BAD$, $\angle CAD$, and $\angle BAE$ in terms of arc lengths.

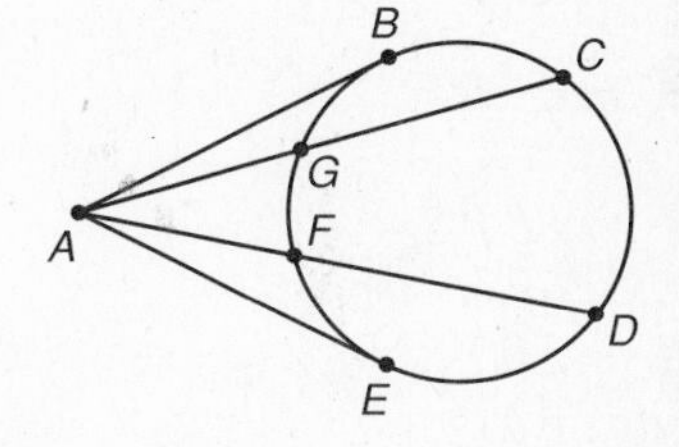

If a tangent and a secant, two tangents, or two secants intersect in the exterior of a circle, then the measure of the angle formed is half the difference of the measures of the intercepted arcs.

1. $m\angle BAC = \frac{1}{2}(m\overset{\frown}{BC} - m\overset{\frown}{BG})$
2. $m\angle BAD = \frac{1}{2}(m\overset{\frown}{BD} - m\overset{\frown}{BF})$
3. $m\angle CAD = \frac{1}{2}(m\overset{\frown}{CD} - m\overset{\frown}{GF})$
4. $m\angle BAE = \frac{1}{2}(m\overset{\frown}{BCE} - m\overset{\frown}{BE})$

c. A floodlight is positioned at point P. If $m\overset{\frown}{QR} = 130^\circ$, what is the measure of the angle of light, $\angle QPR$?

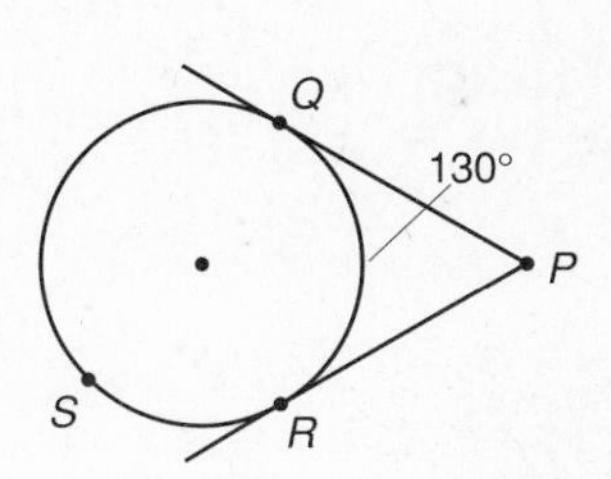

$$\begin{aligned} m\angle QPR &= \tfrac{1}{2}(m\overset{\frown}{QSR} - m\overset{\frown}{QR}) \\ &= \tfrac{1}{2}(230^\circ - 130^\circ) \\ &= \tfrac{1}{2}(100^\circ) \\ &= 50^\circ \end{aligned}$$

Reteach

Chapter 10

Name ______________________

Guidelines: • To find the measures of angles formed by chords, tangents, and secants, you need to know the measures of intercepted arcs and sums or differences of the measures of intercepted arcs.

EXERCISES

In the figure at the right, $m\widehat{SQ} = 114°, m\widehat{PQ} = 130°$. Use the figure at the right to complete each of the following statements.

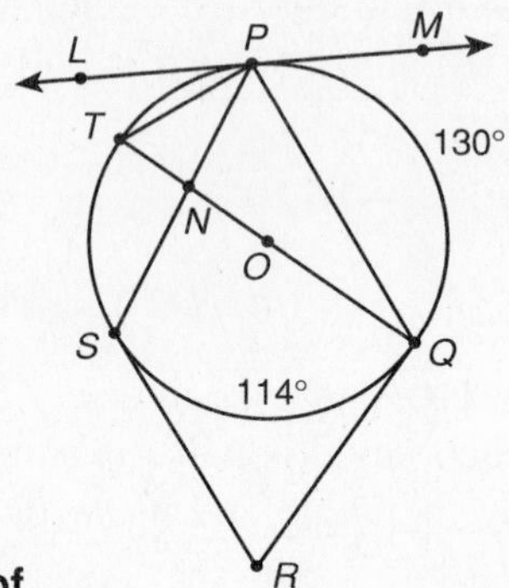

1. $m\angle PNT = \boxed{?}$
2. $m\ \boxed{?} = 65°$
3. $m\angle SRQ = \boxed{?}$
4. $m\ \boxed{?} = 115°$
5. $m\angle SNT = \boxed{?}$
6. $m\ \boxed{?} = 25°$

Use the figure at the right to match each of the following measures of angles and arcs. In the figure, $m\widehat{UP} = 40°$, $m\widehat{QS} = 35°$, $m\widehat{UT} = 45°$.

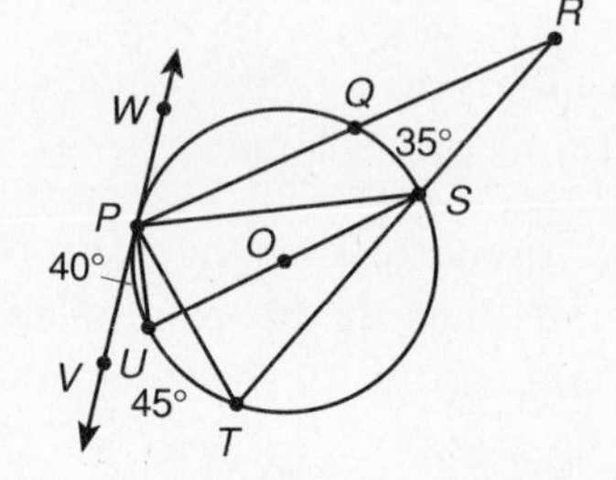

7. $m\angle QRS$	a. 20°
8. $m\widehat{PQ}$	b. 135°
9. $m\angle TPS$	c. 52.5°
10. $m\angle PTS$	d. 70°
11. $m\angle UPV$	e. 25°
12. $m\widehat{ST}$	f. 22.5°
13. $m\angle UST$	g. 105°
14. $m\angle WPQ$	h. 67.5°

Reteach
Chapter 10

Name ____________________

What you should learn:

10.7 How to find the equation of a circle and use equations of circles to solve real-life problems

Correlation to Pupil's Textbook:

Chapter Test (p. 529)
Exercises 17, 18, 20

Examples *Finding Equations of Circles and Using Equations of Circles*

a. The point $(-1, 1)$ is on a circle whose center is $(3, -2)$. Write the standard equation of the circle.

Use the distance formula to find r, the radius.

$r = \sqrt{(x-h)^2 + (y-k)^2}$ — *where* $(x, y) = (-1, 1)$ *and* $(h, k) = (3, -2)$.

$r = \sqrt{(-1-3)^2 + (1-(-2))^2}$ — *Substitute* $x = -1, h = 3, y = 1, k = -2$.

$r = \sqrt{16+9}$ — *Square* -4 *and square* 3.

$r = \sqrt{25}$ — *Simplify.*

$r = 5$ — *Take the positive square root.*

The standard equation of a circle with radius r and center (h, k) is $(x-h)^2 + (y-k)^2 = r^2$. For $(h, k) = (3, -2)$ and $r = 5$, the equation is $(x-3)^2 + (y-(-2))^2 = 5^2$ or $(x-3)^2 + (y+2)^2 = 25$.

b. Assume the radius of the earth is 4000 miles. Find the standard equation for the path of a satellite 20,000 miles above the earth if the origin of the coordinate system is at the center of the earth.

If the radius of the earth is 4000 miles and the distance from the surface of the earth to the satellite is 20,000 miles, then the radius of the orbit is 24,000 miles. Since the center of the circular path is the origin (0, 0), the standard equation is $x^2 + y^2 = (24{,}000)^2$.

Guidelines: To find the equation of a circle that passes through three points

- Draw the perpendicular bisectors to two sides of the triangle formed by the three points.
- The point of intersection of the perpendicular bisectors is the center.
- The distance between the center and any of the given points is the radius.
- The standard equation of a circle with center (h, k) and radius r is $(x-h)^2 + (y-k)^2 = r^2$.

EXERCISES

In Exercises 1–3, determine the center and radius of the circle.

1. $(x+3)^2 + (y-2)^2 = 9$ **2.** $(x-5)^2 + y^2 = \frac{1}{4}$ **3.** $(x+1)^2 + (y-1)^2 = 3$

4. Find an equation of the circle that passes through $P(-2, -5)$, $Q(5, 2)$, and $R(4, 3)$.

Reteach

Chapter 11

Name ______________________________

What you should learn:

11.1	How to find the perimeter of a polygon and the area of a square and rectangle

Correlation to Pupil's Textbook:

Mid-Chapter Self-Test (p. 552)
Exercises 1–5, 9, 12, 15, 16

Chapter Test (p. 583)
Exercises 1–8, 19

Examples — *Finding the Perimeter and the Area of Polygons*

a. Find the perimeter of the polygon $ABCD$ shown at the right.

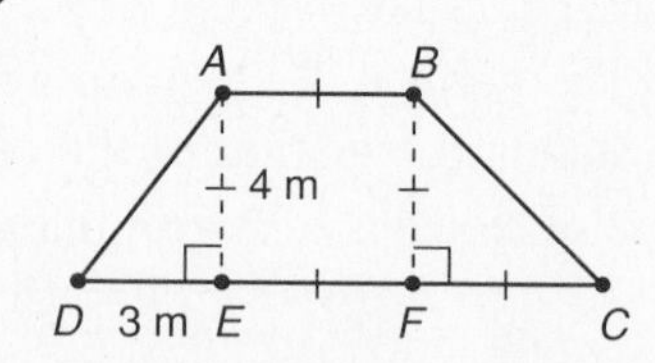

$\overline{AD}$ is the hypotenuse of a right triangle whose legs have lengths 3 m and 4 m; therefore, $AD = 5$ m. $ABFE$ is a square; thus, $AB = BF = EF = 4$ m. Right triangle BFC is isosceles; therefore, $FC = 4$ m and hypotenuse $BC = 4\sqrt{2}$ m. The perimeter of polygon $ABCD$ is the sum of the lengths of the sides.

Perimeter	$= AB + BC + CD + AD$	*Formula for perimeter*
	$= 4 + 4\sqrt{2} + 11 + 5$	*Substitute lengths.*
	$= 20 + 4\sqrt{2}$	*Simplify.*

The perimeter is about 25.66 meters.

b. In the figure shown at the right, $ABCD$ is a square and $DGFE$ is a rectangle. Find the area of the entire region.

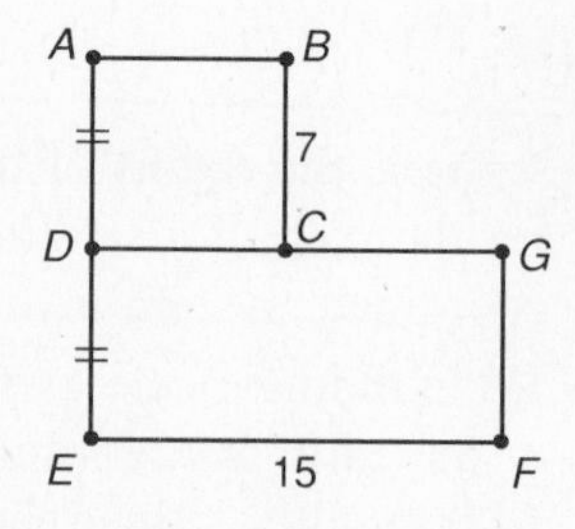

The area of square $ABCD$ is the square of the length of its side, or $A = 7^2 = 49$ square units. The area of rectangle $DGFE$ is the product of its base and height, or $A = 15(7) = 105$ square units. By the Area Addition Postulate, the area of a region is the sum of all its nonoverlapping parts. Adding the areas of the square $ABCD$ and the rectangle $DGFE$, the area of the region is $A = 49 + 105 = 154$ square units.

Guidelines:

- *Perimeter* is a linear measurement while *area* is measured in square units.

EXERCISES

1. Find the area of the shaded region.

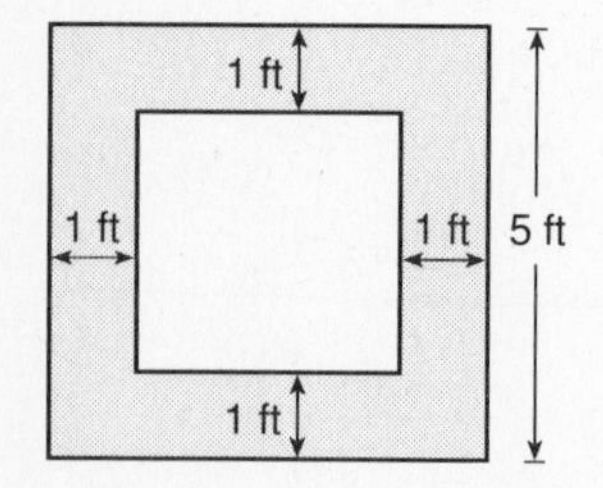

2. Find the perimeter of the parallelogram $ABCD$.

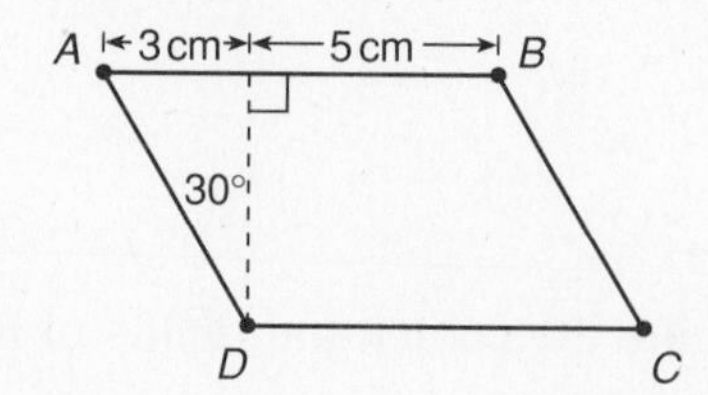

Reteach
Chapter 11

Name ____________________

What you should learn:

11.2 How to find the area of a parallelogram and the area of a triangle

Correlation to Pupil's Textbook:

Mid-Chapter Self-Test (p. 552)
Exercises 5, 6, 11, 13, 17, 19

Chapter Test (p. 583)
Exercises 5, 6

Examples *Finding the Area of a Parallelogram and the Area of a Triangle*

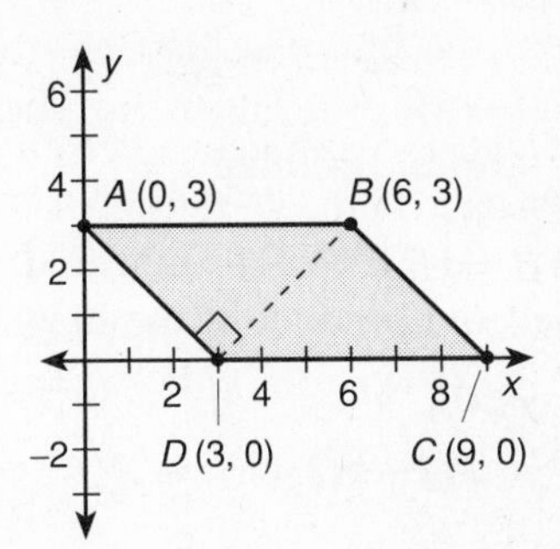

a. Find the area of $\square ABCD$ using two different sets of measurements.

1. Choose $\overline{DC}$ as the base. Then $b = 6$ and $h = 3$. The area is $A = bh = (6)(3) = 18$ square units.
2. Choose $\overline{AD}$ as the base. Use the Pythagorean Theorem to solve for $AD = b$ and $BD = h$. Then $b = 3\sqrt{2}$ and $h = 3\sqrt{2}$. The area is $A = bh = (3\sqrt{2})(3\sqrt{2}) = 18$ square units.

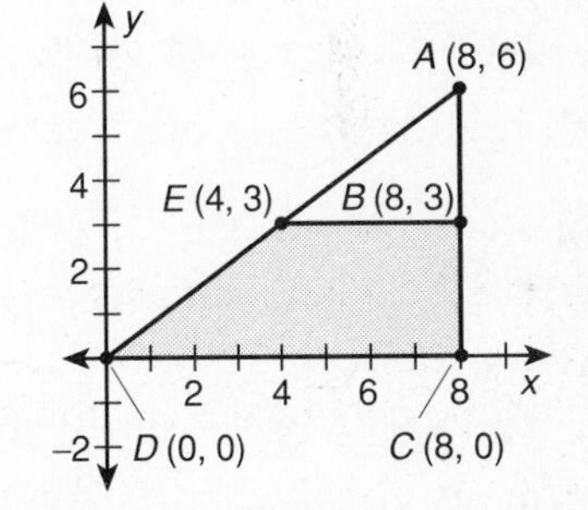

b. Find the area of the shaded region $BCDE$ in the figure shown at the right.

Using the Area Addition Postulate, find the area of the shaded region $BCDE$ by subtracting the area of $\triangle ABE$ from the area of $\triangle ACD$.
Area of $\triangle ACD = \frac{1}{2}bh = \frac{1}{2}(8)(6) = 24$ square units and
Area of $\triangle ABE = \frac{1}{2}bh = \frac{1}{2}(4)(3) = 6$ square units.
Therefore, the area of the shaded region $BCDE$ is 18 square units.

Guidelines:

- A ratio of the areas of two polygons can be used to find the probability that an event will occur.

EXERCISES

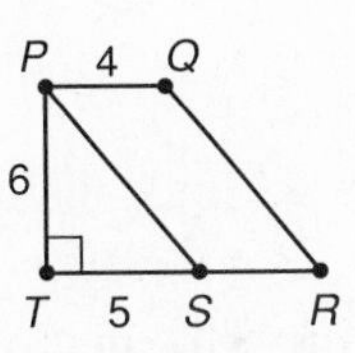

1. Find the area of $\square PQRS$ and $\triangle PST$.

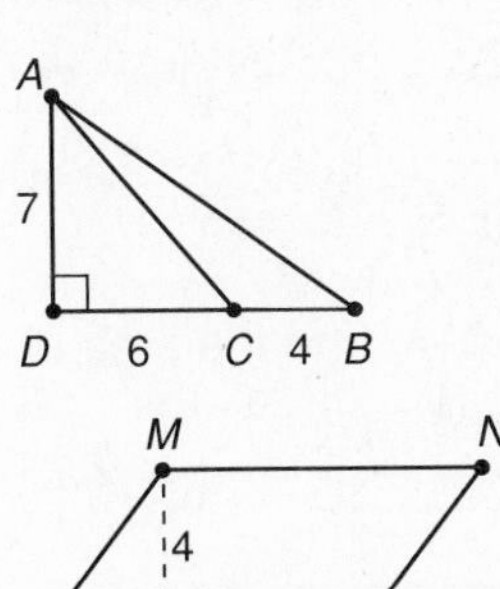

2. a. Find the area of $\triangle ABC$ using the Area Addition Postulate.
b. Find the area of $\triangle ABC$ using the Area of a Triangle Theorem.
(Show your work for each calculation.)

3. Suppose that an event is equally likely to occur anywhere in $\square MNOQ$, shown at the right. What is the probability that it will occur in $\triangle MPQ$?

Name ____________________

What you should learn:

11.3	How to find the area of a trapezoid and the area of a quadrilateral whose diagonals are perpendicular

Correlation to Pupil's Textbook:

Mid-Chapter Self-Test (p. 552) Exercises 7, 8, 10, 14, 18, 20

Chapter Test (p. 583) Exercise 8

Examples *Finding the Area of a Trapezoid and the Areas of Other Quadrilaterals*

a. Use the figure at the right to find the area of trapezoid $RSTQ$ using the Area of a Trapezoid Theorem. Verify your answer using the Area Addition Postulate.

By the Area of a Trapezoid Theorem, the area of trapezoid $RSTQ$ is half the product of the height and the sum of the bases, or $A = \frac{1}{2}h(b_1 + b_2)$. For $h = 3$, $b_1 = 5$, and $b_2 = 10$,

$$A = \tfrac{1}{2}(3)(15) = 22.5 \text{ square units.}$$

To verify this answer using the Area Addition Postulate,

$$\text{Area } \triangle PQT + \text{ Area Trapezoid } RSTQ = \text{ Area } \triangle PRS$$
$$\tfrac{1}{2}(5)(3) + \text{ Area Trapezoid } RSTQ = \tfrac{1}{2}(10)(6)$$
$$\tfrac{15}{2} + \text{ Area Trapezoid } RSTQ = 30$$
$$\text{Area Trapezoid } RSTQ = \tfrac{45}{2} = 22.5 \text{ square units}$$

b. Find the area of the quadrilateral shown at the right.

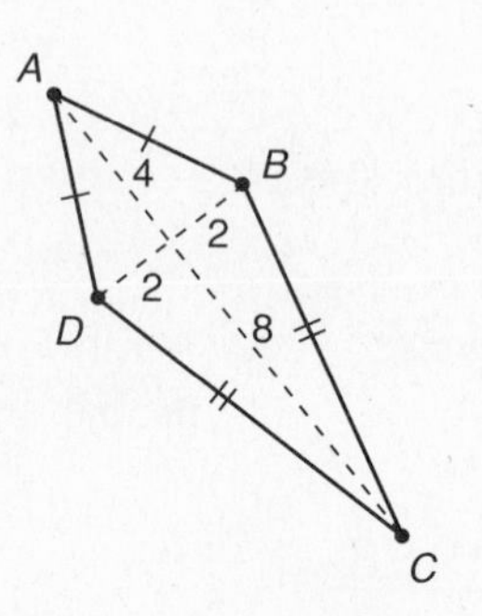

The quadrilateral is a kite, therefore, its diagonals are perpendicular. If the diagonals of a quadrilateral are perpendicular, then the area of the quadrilateral is half the product of the lengths of the diagonals, or $A = \frac{1}{2}d_1d_2$. Let $DB = d_1$ and $AC = d_2$. Then $A = \frac{1}{2}(4)(12)$ $= 24$ square units.

Guidelines:

- Since the diagonals of a rhombus are perpendicular, the area of a rhombus is half the product of the diagonals.
- The area of a trapezoid is half the product of the height and the sum of the bases, or $A = \frac{1}{2}h(b_1 + b_2)$.

EXERCISES

In Exercises 1–4, find the area of the quadrilateral.

1.

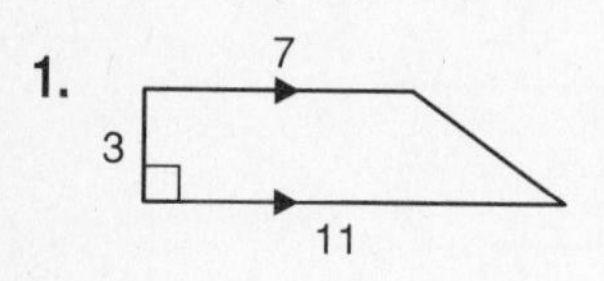

2.

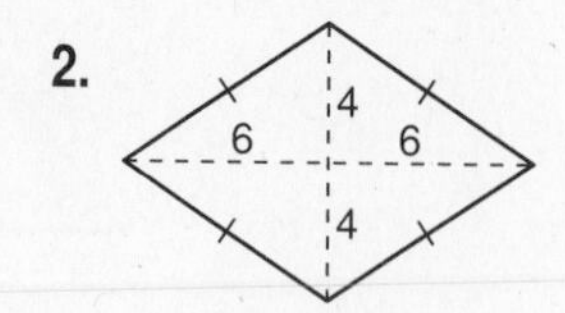

3.

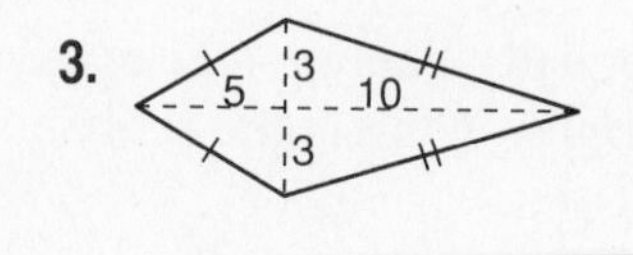

4.

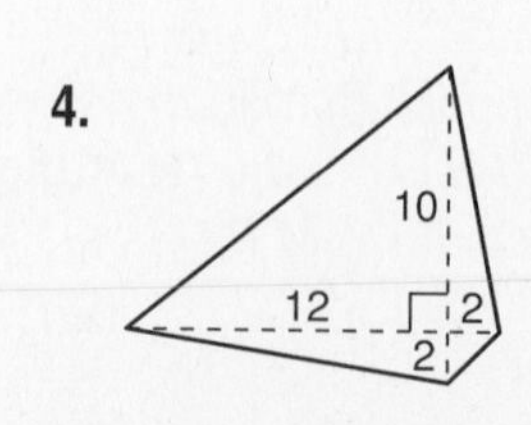

Reteach
Chapter 11

Name ____________________

What you should learn:

11.4 How to find the area of an equilateral triangle and the area of a regular polygon

Correlation to Pupil's Textbook:

Chapter Test (p. 583)
Exercise 7

Examples *Finding the Area of an Equilateral Triangle and a Regular Polygon*

a. Find the area of $\triangle ABD$, shown at the right, using the Area of an Equilateral Triangle Theorem. Verify your answer using the Area of a Triangle Theorem.

By the Area of an Equilateral Triangle Theorem, the area is one-fourth the square of the length of the side times $\sqrt{3}$, or $A = \frac{1}{4}\sqrt{3}(10)^2$. The area is $25\sqrt{3}$ square units, or about 43.30 square units.

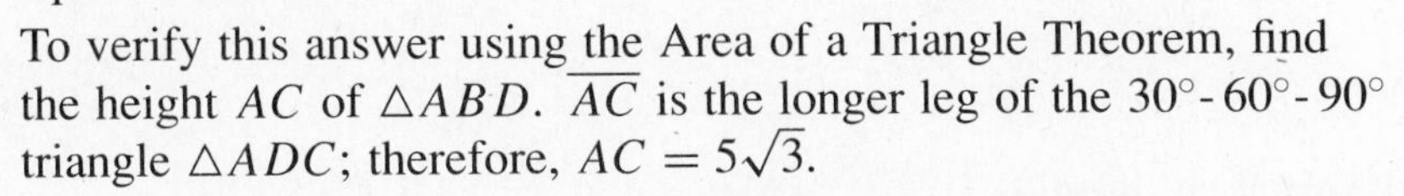

To verify this answer using the Area of a Triangle Theorem, find the height AC of $\triangle ABD$. $\overline{AC}$ is the longer leg of the 30°-60°-90° triangle $\triangle ADC$; therefore, $AC = 5\sqrt{3}$.

Area of $\triangle ABD = \frac{1}{2}bh = \frac{1}{2}(10)(5\sqrt{3}) = 25\sqrt{3}$ square units, or about 43.30 square units.

b. Find the area of the regular hexagon $ABCDEF$, shown at the right.

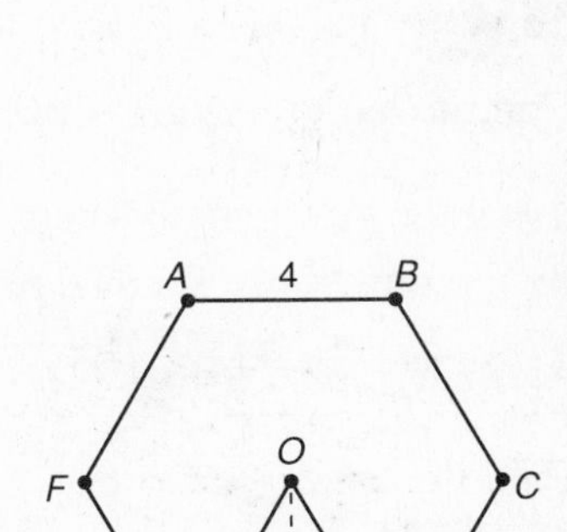

The length of $\overline{AB}$ is 4 inches. To use the Area of a Regular Polygon Theorem, you must solve for the length of the apothem, $\overline{OG}$. The measure of $\angle EOD$ is $\frac{1}{6}(360°)$ or 60°. $\triangle OGD$ is a 30°-60°-90° right triangle with $GD = 2$ and $OG = 2\sqrt{3}$. The perimeter of the hexagon is $6(4) = 24$ inches. The area of a regular polygon is half the product of the apothem, a, and the perimeter, P, or $A = \frac{1}{2}aP = \frac{1}{2}(2\sqrt{3})(24)$. The area of hexagon $ABCDEF$ is $24\sqrt{3}$ square inches, or about 41.57 square inches.

Guidelines:

- The apothem of a regular polygon is the distance between the center and a side.
- A central angle of a regular polygon is an angle whose vertex is the center and whose sides contain two consecutive vertices of the polygon.

EXERCISES

In Exercises 1–3, find the perimeter and area of each regular polygon.

1. (square, 6)

2.

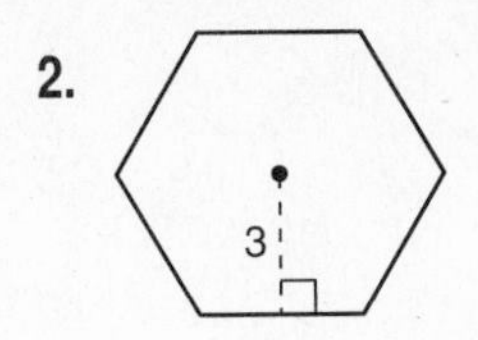

3.

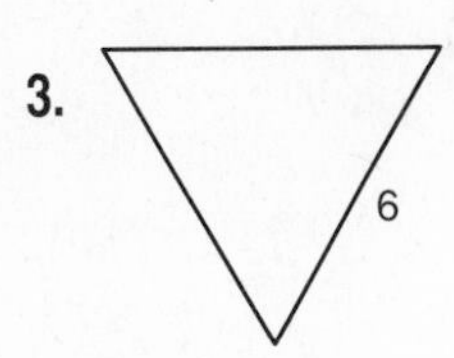

4. Explain another method for solving Exercise 1.

Reteach
Chapter 11

Name ____________________

What you should learn:

11.5	How to find the circumference of a circle and the length of an arc

Correlation to Pupil's Textbook:

Chapter Test (p. 583)
Exercises 11, 12, 17

Examples *Finding the Circumference of a Circle and the Length of an Arc*

a. Find the circumference of the circle shown at the right.

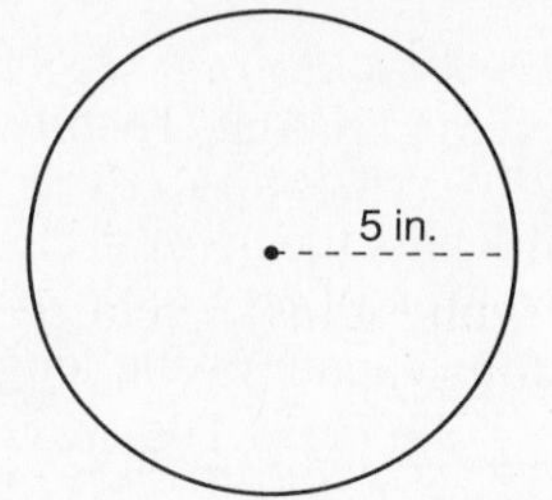

Using the Circumference of a Circle Theorem, the circumference is $C = \pi d$ or $C = 2\pi r$. For $r = 5$ in., $C = 2\pi(5\text{in.}) = 10\pi$ inches. The distance around the circle is about 31.42 inches.

b. Find the length of $\widehat{CB}$, shown in the circle at the right.

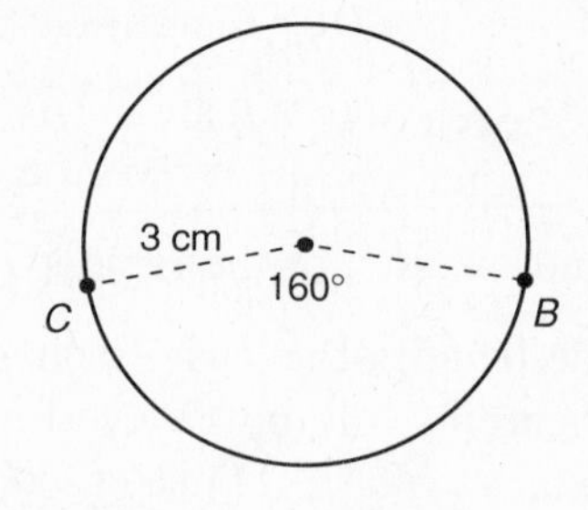

Using the Arc Length Theorem, the ratio of the length of arc $\widehat{CB}$ to the circumference is equal to the ratio of $m\widehat{CB}$ to 360°.

$$\frac{\text{Arc length of } \widehat{CB}}{2\pi r} = \frac{m\widehat{CB}}{360^\circ} \text{ where } m\widehat{CB} = 160^\circ.$$

$$\text{Arc length of } \widehat{CB} = 2\pi r\left(\frac{160^\circ}{360^\circ}\right) = 2\pi(3)\left(\frac{4}{9}\right) \approx 8.38 \text{ centimeters.}$$

Guidelines:

- An arc length is a portion of the circumference of a circle.
- The measure of an arc is given in degrees and the length of an arc is given in linear units such as inches or centimeters.

EXERCISES

In Exercises 1–3, find the indicated measure.

1.

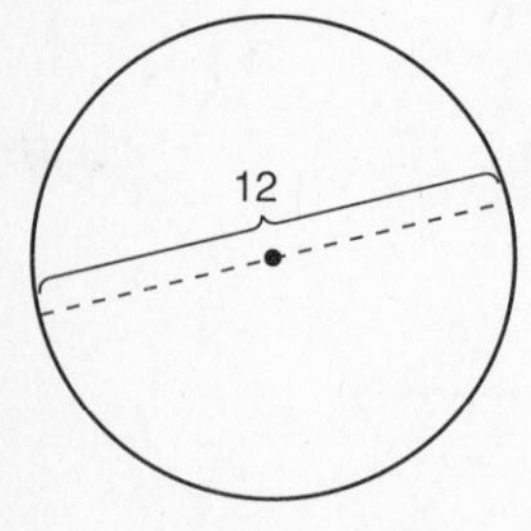

Circumference

2.

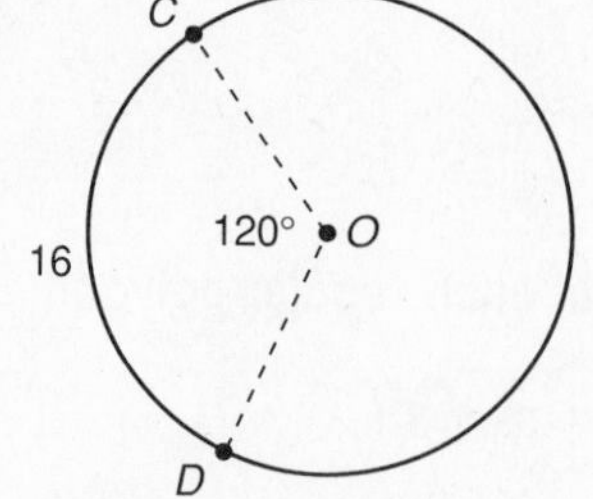

Radius

3.

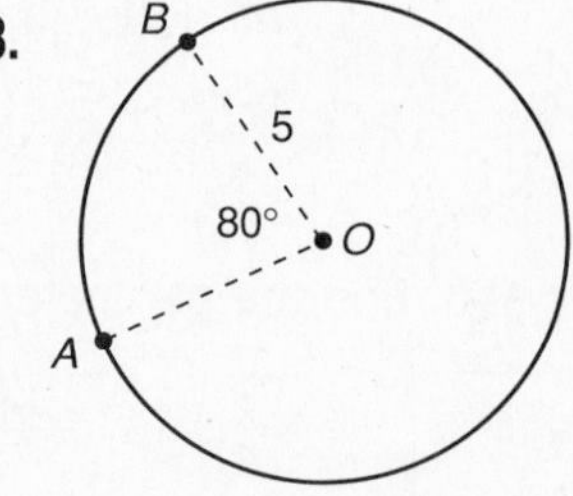

Length of $\widehat{AB}$

Reteach
Chapter 11

Name ______________________________

What you should learn:

11.6	How to find the area of a circle and areas of regions of circles

Correlation to Pupil's Textbook:

Chapter Test (p. 583)
Exercises 9, 10, 13–16, 18

Examples *Finding the Area of a Circle and Areas of Regions of Circles*

a. Find the area of the shaded region, shown at the right.

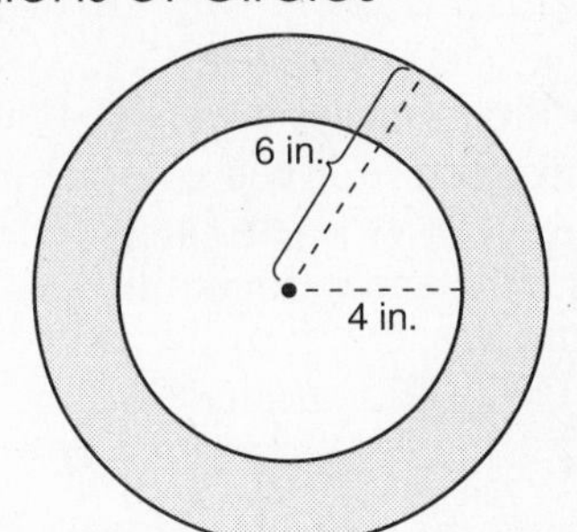

Using the Area Addition Postulate, the area of the circle with radius $= 4$ in. plus the area of the shaded region equals the area of the circle with radius $= 6$ in. By the Area of a Circle Theorem, the area of a circle is π times the square of the radius, or $A = \pi r^2$. Then,

$$\pi(4)^2 + \text{ Area of shaded region} = \pi(6)^2$$

$$16\pi + \text{ Area of shaded region} = 36\pi$$

$$\text{Area of shaded region} = 20\pi \approx 62.83 \text{ square inches.}$$

b. Find the area of the sector XOY. Find the area of the minor segment bounded by $\overline{XY}$ and $\widehat{XY}$. Use the figure at the right.

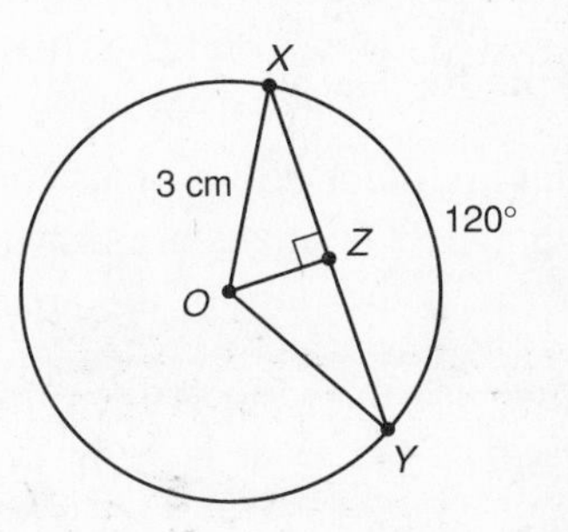

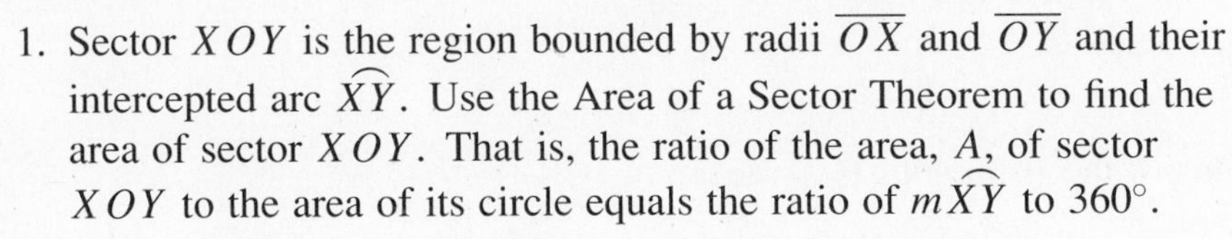

1. Sector XOY is the region bounded by radii $\overline{OX}$ and $\overline{OY}$ and their intercepted arc $\widehat{XY}$. Use the Area of a Sector Theorem to find the area of sector XOY. That is, the ratio of the area, A, of sector XOY to the area of its circle equals the ratio of $m\widehat{XY}$ to 360°.

   $\dfrac{A}{\pi r^2} = \dfrac{m\widehat{XY}}{360^\circ}$ or $A = \dfrac{m\widehat{XY}}{360^\circ}\pi r^2 = \dfrac{120^\circ}{360^\circ}(\pi 3^2) = 3\pi \text{ cm}^2$
   ≈ 9.42 square centimeters.

2. To find the area of the minor segment, subtract the area of $\triangle XOY$ from the area of the sector. Solve the 30°-60°-90° triangle XOZ for $OZ = \frac{3}{2}$ and $XZ = \frac{3}{2}\sqrt{3}$. Double XZ to determine $\overline{XY}$, the base of $\triangle XOY$. Then the base is $3\sqrt{3}$ and the height is $\frac{3}{2}$. The area of $\triangle XOY = \frac{1}{2}bh = \frac{1}{2}(3\sqrt{3})\left(\frac{3}{2}\right) = \frac{9}{4}\sqrt{3} \approx 3.9$ square centimeters. The area of the minor segment is $\left(3\pi - \frac{9}{4}\sqrt{3}\right)$ cm^2 or ≈ 5.53 square centimeters.

Guidelines:

- A segment is called a minor segment if the measure of the intercepted arc is less than 180°.

EXERCISES

In Exercises 1–3, find the area of the shaded region.

1.

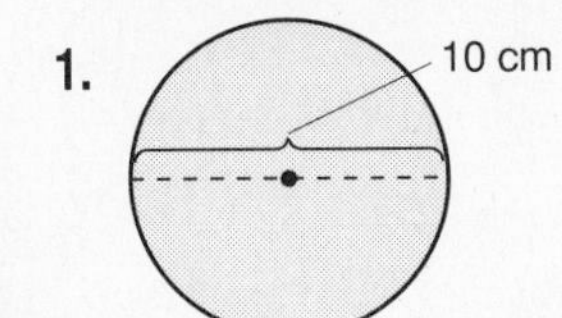

2.

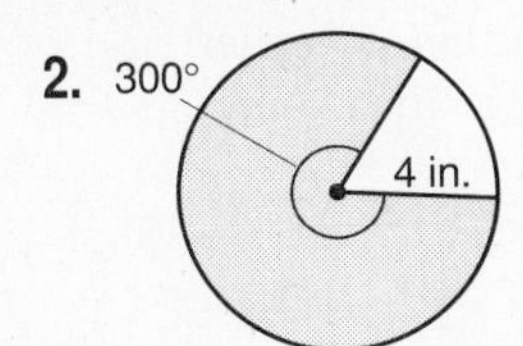

3.

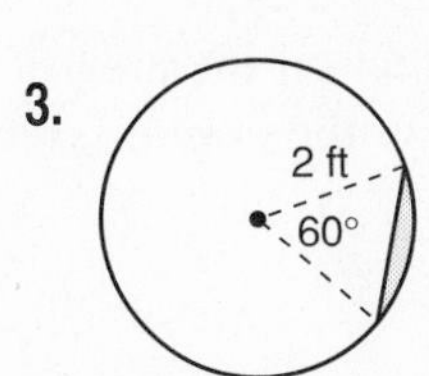

Reteach
Chapter 11

Name ______________________________

What you should learn:

11.7 How to find areas of similar polygons and use areas of similar polygons to solve real-life problems

Correlation to Pupil's Textbook:

Chapter Test (p. 583)
Exercise 20

Examples *Finding Areas of Similar Polygons and Using the Areas in Real Life*

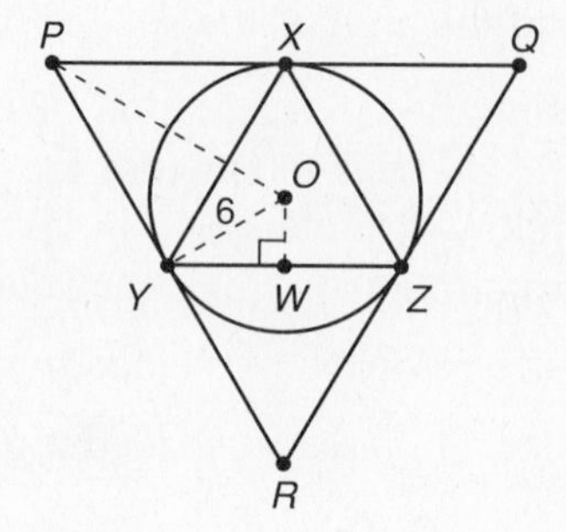

a. In the figure at the right, find the ratio of the areas of the inscribed and circumscribed equilateral triangles $\triangle XYZ$ and $\triangle PQR$. The length of $\overline{OY}$ is 6 inches.

Find the length of the side of each equilateral triangle using $30°$-$60°$-$90°$ triangles. In $\triangle OWY$, $YW = 3\sqrt{3}$; therefore $YZ = 6\sqrt{3}$. In $\triangle OPY$, $PY = 6\sqrt{3}$; therefore $PR = 12\sqrt{3}$. Using the Areas of Similar Polygons Theorem, two equilateral triangles are similar polygons with corresponding sides in the ratio of $6\sqrt{3} : 12\sqrt{3}$ or 1:2. Then the ratio of the areas is $1^2 : 2^2$ or 1:4.

b. An architect draws plans for the construction of a new rectangular garage. The plans are drawn to the scale of $\frac{1}{4}$ inch to 1 foot, or 1:48. What is the ratio of the areas of the drawing and the actual garage?

Because the ratio of the dimensions of the rectangles is 1:48, the ratio of the areas is $1:48^2$ or 1:2304.

Guidelines:

- If two polygons are similar, the ratio of the perimeters of the polygons is equal to the ratio of the corresponding sides.
- If two polygons are similar, the ratio of the areas of the polygons is equal to the square of the ratio of the corresponding sides.

EXERCISES

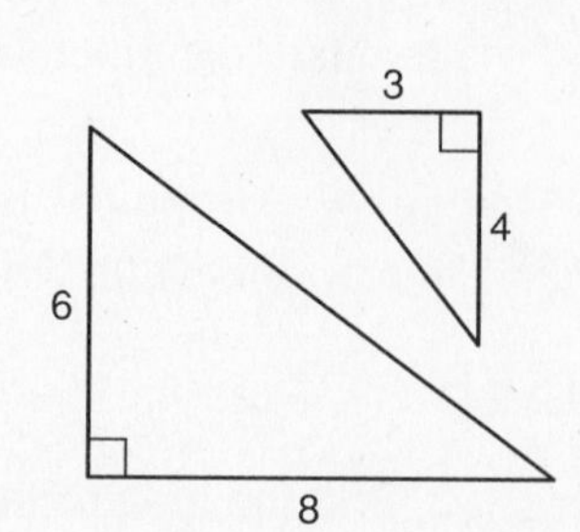

1. The figures shown at the right are similar. Find the ratio of their perimeters and the ratio of their areas.

2. What is the effect on the area of a circle if its radius is tripled? What happens to the circumference?

Reteach

Chapter 12

Name ______________________________

What you should learn:

12.1 How to identify solids that are polyhedrons and use polyhedrons to solve real-life problems

Correlation to Pupil's Textbook:

Mid-Chapter Self-Test (p. 606)	**Chapter Test (p. 637)**
Exercises 2, 10, 18–20	Exercises 1, 2

Examples *Identifying Polyhedrons and Using Polyhedrons in Real Life*

a. Two solids are shown at the right. Explain why only one solid is a polyhedron. Count the faces, vertices, and edges of the solid that is a polyhedron.

(a)

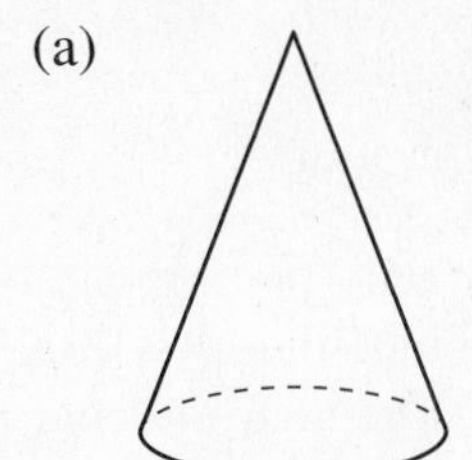

(b)

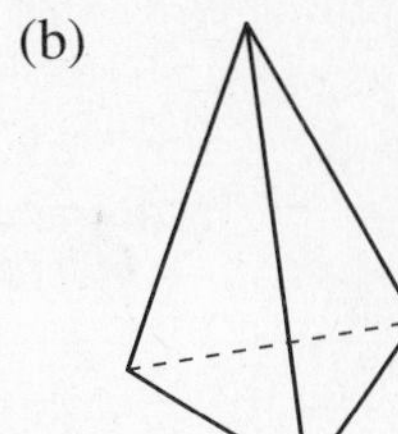

1. Figure (a) is not a polyhedron because it is not a solid bounded by polygons. Figure (b) is a polyhedron. The faces are triangles that enclose a single region of space.
2. Polyhedron (b) has 4 triangular faces, 6 edges (segments formed by the intersection of faces), and 4 vertices (points where three or more edges meet).

b. Use Euler's Theorem to relate the number of faces (F), vertices (V), and edges (E) of the polyhedron shown at the right.

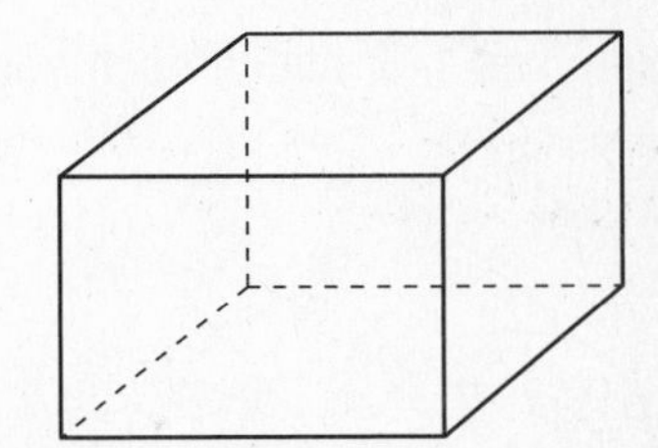

The polyhedron has 6 faces, 8 vertices, and 12 edges. By Euler's Theorem, the number of faces (F), vertices (V), and edges (E) of a polyhedron is related by $F + V = E + 2$, or $6 + 8 = 12 + 2$.

c. Two polyhedrons are shown at the right. Explain why only one polyhedron is convex and determine whether the convex polyhedron is regular.

(a)

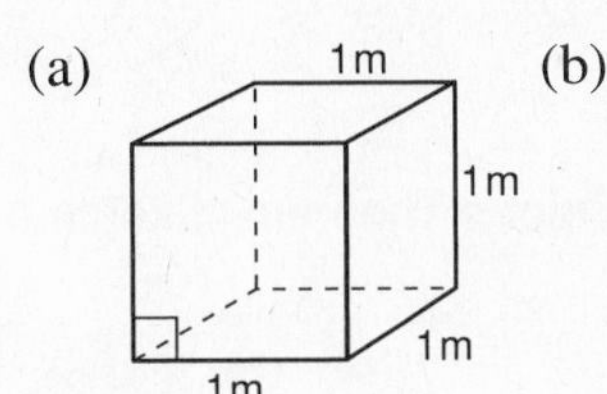

(b)

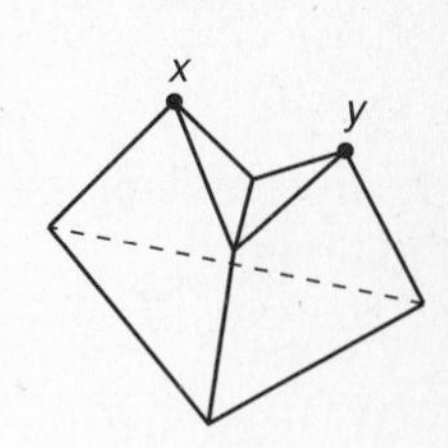

1. A polyhedron is convex if any two points on its surface (all points on its faces) can be connected by a line segment that lies entirely inside or on the polyhedron. Polyhedron (b) is not convex because the segment connecting the points X, and Y lies outside the polyhedron. Polyhedron (a) is convex because any two points on its surface connected by a line segment will lie entirely inside or on the polyhedron.
2. A polyhedron is regular if all its faces are congruent regular polygons, and the same number of faces meet at each vertex in exactly the same way. Polyhedron (a) is regular.

Name ______________________________

d. Buckminster Fuller created geodesic domes. Geodesic domes are built using a combination of tetrahedrons and octahedrons. Count the number of faces, vertices, and edges of the octahedron. Verify you answer by using Euler's Theorem.

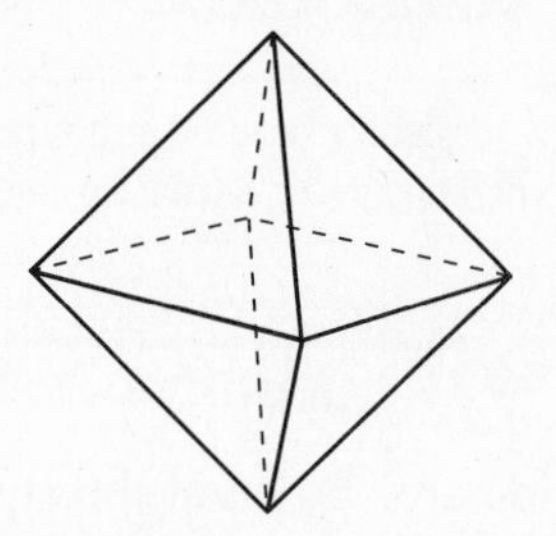

The octahedron has 8 faces, 6 vertices, and 12 edges. By Euler's Theorem, $F + V = E + 2$ or $8 + 6 = 12 + 2$.

Guidelines:

- There are only five kinds of *regular* polyhedrons: a regular tetrahedron (4 faces), a cube (6 faces), a regular octahedron (8 faces), a regular dodecahedron (12 faces), and a regular icosahedron (20 faces).
- A *semiregular* polyhedron is one whose faces are more than one type of regular polygon and whose vertices are all formed in exactly the same way.

EXERCISES

In Exercises 1–3, match each solid with the correct description.

1. 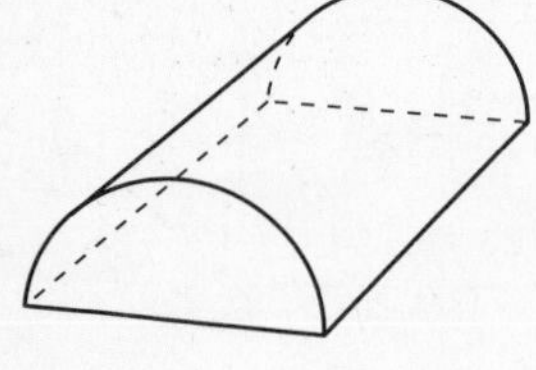

2. 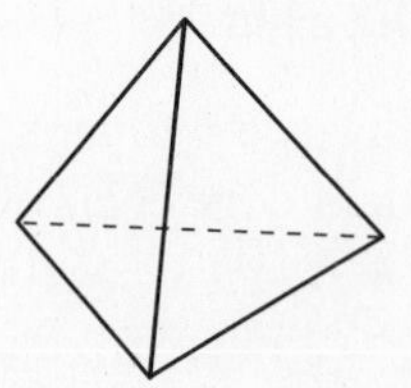

3. 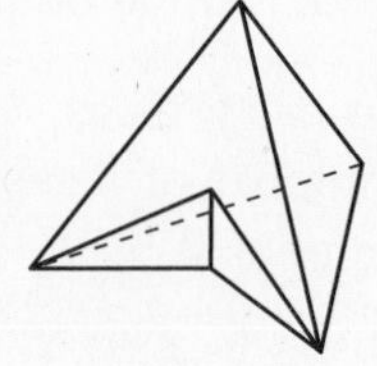

(a) Regular polyhedron (b) Not a polyhedron (c) Not a convex polyhedron

In Exercises 4–6, use Euler's Theorem. Sketch a figure that satisfies the conditions.

4. Faces: 8
Vertices: [?]
Edges: 12

5. Faces: 4
Vertices: 4
Edges: [?]

6. Faces: [?]
Vertices: 8
Edges: 12

In Exercises 7–10, identify the regular polyhedron.

7. 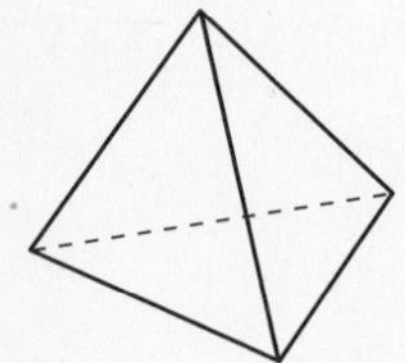

8. 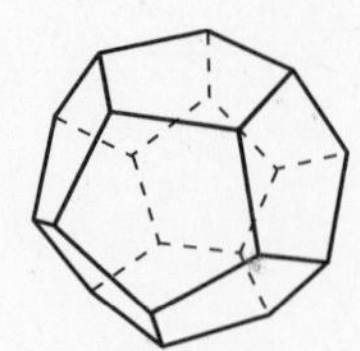

9. 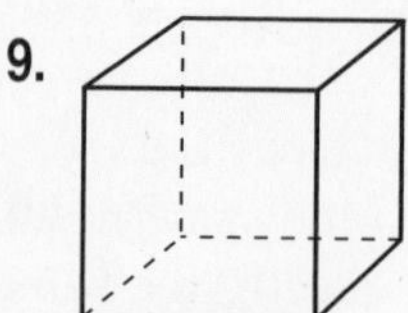

10. 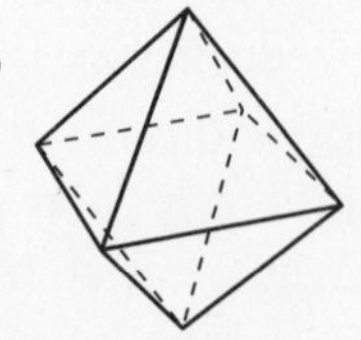

Reteach
Chapter 12

Name ______________________________

What you should learn:

12.2 How to find the surface area of a prism

Correlation to Pupil's Textbook:

Mid-Chapter Self-Test (p. 606) Exercises 16, 17

Chapter Test (p. 637) Exercises 3, 8

Examples *Finding the Surface Area of a Prism*

a. Match the following prism definitions with the correct term.

		Answers:
1. bases of a prism	a. parallel, congruent faces	1. a
2. lateral faces	b. perpendicular distance between the bases	2. g
3. lateral edges	c. prism with lateral edges that are oblique to the bases	3. i
4. altitude or height	d. prism with rectangular bases	4. b
5. right prism	e. sum of the areas of the faces	5. j
6. oblique prism	f. length of the oblique lateral edges	6. c
7. slant height	g. prism faces that are parallelograms formed by connecting corresponding vertices of bases	7. f
8. rectangular prism	h. a pattern that can be cut and folded to form a polyhedron	8. d
9. triangular prism	i. segments connecting corresponding vertices of bases	9. k
10. surface area	j. prism whose lateral edges are perpendicular to both bases	10. e
11. net	k. prism with triangular bases	11. h

b. Find the surface area of the right prism shown at the right, in which the bases of the prism are equilateral triangles.

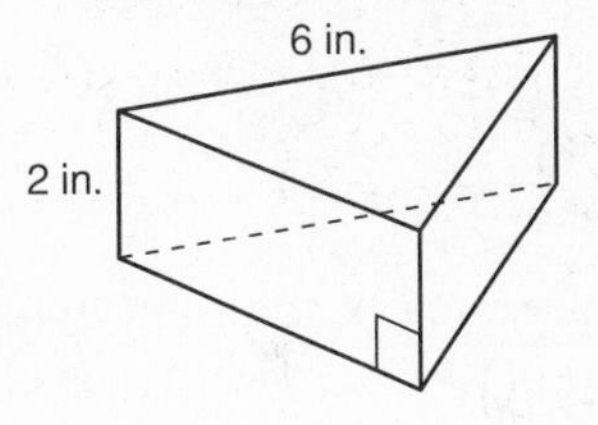

The surface area, S, of a right prism is $S = 2B + Ph$, where B is the area of a base, P is the perimeter of a base, and h is the height. Solve for B using the formula for the area of an equilateral triangle.

$$B = \frac{1}{4}s^2\sqrt{3} = \frac{1}{4}(6)^2\sqrt{3} = 9\sqrt{3} \text{ square inches}$$

Since $P = 18$ inches and $h = 2$ inches, $S = 2(9\sqrt{3}) + 18(2) = (18\sqrt{3} + 36)$ square inches. The surface area of the right prism is about 67.18 square inches.

Guidelines:

- The surface area of a right prism is $S = 2B + Ph$, where B is the area of a base, P is the perimeter of a base, and h is the height.

EXERCISES

1. Find the surface area of the right rectangular prism.

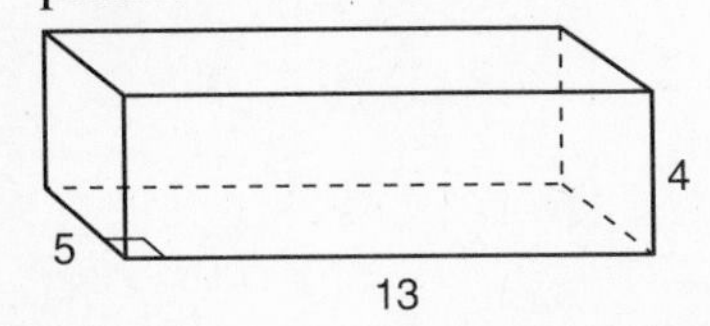

2. Find the surface area of the right triangular prism.

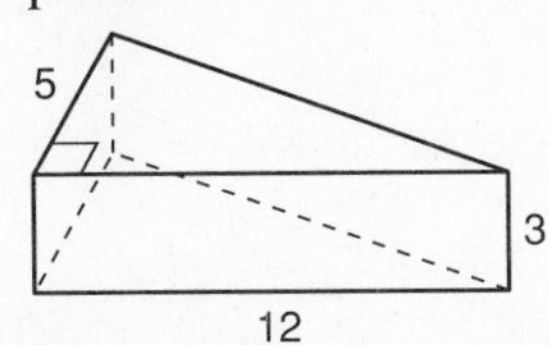

Reteach

Chapter 12

Name ______________________

What you should learn:

12.2 How to find the surface area of a cylinder

Correlation to Pupil's Textbook:

Mid-Chapter Self-Test (p. 606) Exercises 3–5, 13

Chapter Test (p. 637) Exercise 4

Examples *Finding the Surface Area of a Cylinder*

a. Complete each of the following cylinder definitions.

1. A solid with congruent circular bases that lie in parallel planes is a [?].
2. The perpendicular distance between the bases is the [?] or [?].
3. The area of the curved surface is the [?] area.
4. In a [?] cylinder, the segment joining the centers of its bases is perpendicular to its bases.

Answers:

1. cylinder
2. altitude, height
3. lateral
4. right

b. Find the surface area of the right cylinder shown at the right.

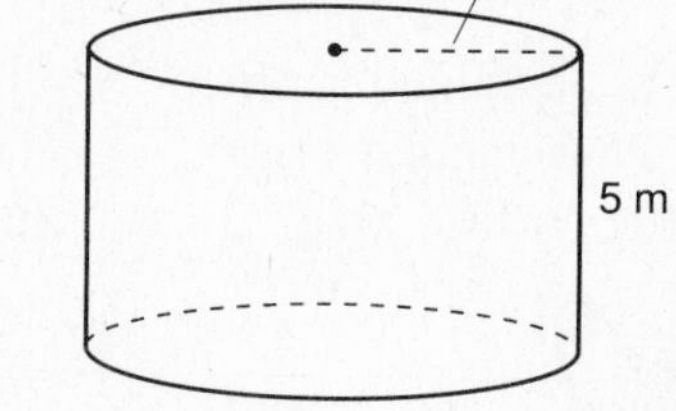

The surface area, S, of a right cylinder is $S = 2B + Ch = 2\pi r^2 + 2\pi rh$, where B is the area of a base, C is the circumference of a base, r is the radius of a base, and h is the height.

$S = 2\pi r^2 + 2\pi rh$	*Formula for surface area*
$= 2\pi(4^2) + 2\pi(4)(5)$	*Substitute for r and h.*
$= 32\pi + 40\pi$	*Simplify.*
$= 72\pi$ m^2	*Simplify.*

The surface area of the right cylinder is about 226.19 square meters.

Guidelines:

- The surface area of a right cylinder is $S = 2B + Ch = 2\pi r^2 + 2\pi rh$, where B is the area of a base, C is the circumference of a base, r is the radius of a base, and h is the height.

EXERCISES

1. Find the surface area of the right cylinder.

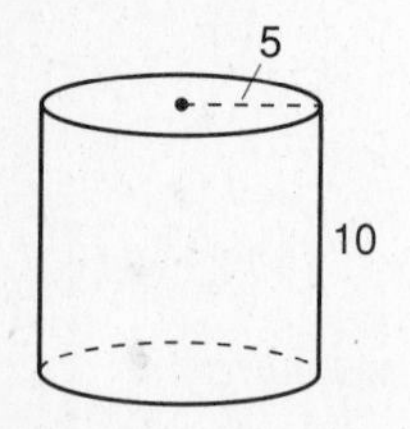

2. Find the lateral area, $2\pi rh$, of the right cylinder.

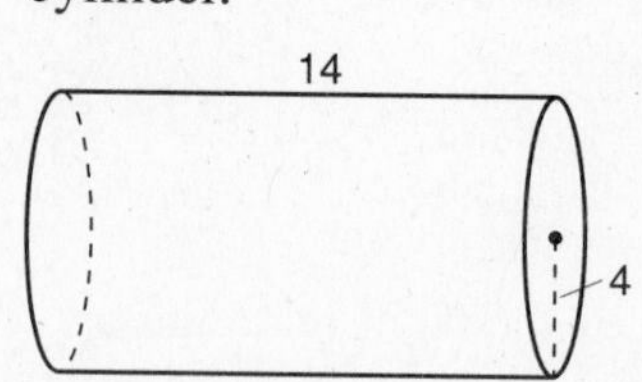

3. Explain the role of C in the formula $S = 2B + Ch$ and compare it to P in the formula $S = 2B + Ph$.

Reteach

Chapter 12

Name ______________________________

What you should learn:

12.3 How to find the surface area of a pyramid

Correlation to Pupil's Textbook:

Mid-Chapter Self-Test (p. 606)
Exercises 2, 6, 7, 9, 12, 15

Chapter Test (p. 637)
Exercises 5, 20

Examples *Finding the Surface Area of a Pyramid*

a. Match the following pyramid definitions with the correct term.

1. a polyhedron in which the base is a polygon and the lateral faces are triangles with a common vertex
2. the intersection of two lateral faces
3. the perpendicular distance between the base and the vertex
4. the intersection of the base and a lateral face
5. a pyramid whose base is a regular polygon and the segment from the vertex to the center of the base is perpendicular to the base

a. lateral edge
b. regular pyramid
c. base edge
d. altitude or height
e. pyramid

Answers: 1. e 2. a 3. d 4. c 5. b

b. Find the surface area of the regular pyramid shown at the right. The base of the pyramid is a regular hexagon with apothem $a = \frac{9}{2}\sqrt{3}$ ft, $s = 9$ ft, and slant height $\ell = 12$ ft.

The surface area, S, of a regular pyramid is $S = B + \frac{1}{2}P\ell$, where B is the area of the base, P is the perimeter of the base, and ℓ is the slant height. Find the area of the base, B, using the formula for the area of a regular hexagon, $B = \frac{1}{2}aP$. Since $a = \frac{9}{2}\sqrt{3}$ and $P = 54$, $B = \frac{1}{2}\left(\frac{9}{2}\sqrt{3}\right)(54) = 121.5\sqrt{3}$ square feet. Then,

$S = B + \frac{1}{2}P\ell$ *Formula for surface area*

$= 121.5\sqrt{3} + \frac{1}{2}(54)(12)$ *Substitute for B, P, and ℓ.*

$= \left(121.5\sqrt{3} + 324\right)\ \text{ft}^2$ *Simplify.*

The surface area of the regular pyramid is about 534.44 square feet.

- In a nonregular pyramid, the altitudes of the lateral faces vary in length, so the pyramid has no slant height.

EXERCISES

1. Find the surface area of a regular pyramid with a square base, as shown below.

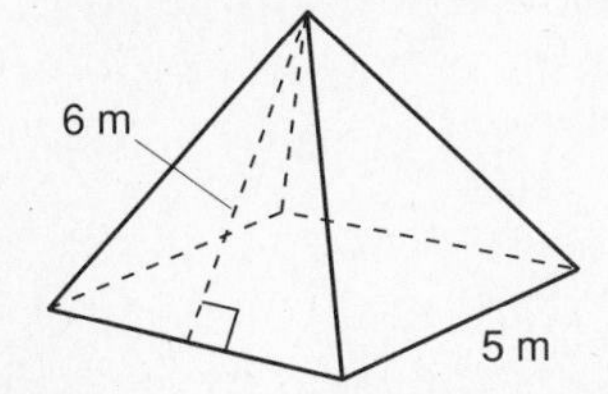

2. Find the surface area of a regular pyramid with a triangular base, as shown below.

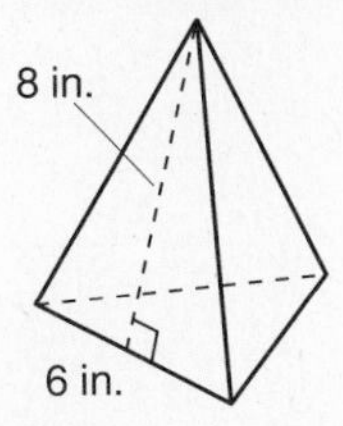

Reteach
Chapter 12

Name ______________________

What you should learn:

12.3 How to find the surface area of a cone

Correlation to Pupil's Textbook:

Mid-Chapter Self-Test (p. 606)
Exercises 1, 8, 11, 14

Chapter Test (p. 637)
Exercise 6

Examples *Finding the Surface Area of a Cone*

a. Complete each of the following circular cone definitions.

1. A solid that has a circular base and a vertex that is not in the same plane as the base is a [?].
2. The [?] surface consists of all segments that connect the vertex with points on the edge of the base.
3. The perpendicular distance between the vertex and the plane that contains the base is the [?] or [?].
4. A [?] cone is one in which the vertex lies directly above the center of the base.
5. The distance between the vertex and a point on the edge of the base is the [?] height.

Answers:

1. circular cone
2. lateral
3. altitude, height
4. right
5. slant

b. Find the surface area of the right cone shown at the right in which the radius of the base is 4 feet and the slant height is 5 feet.

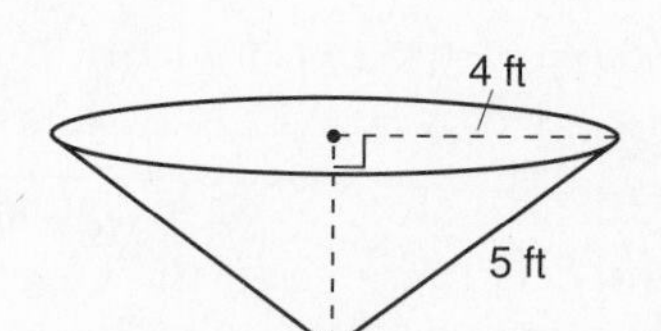

The surface area, S, of a right cone is $S = \pi r^2 + \pi r \ell$, where r is the radius of the base and ℓ is the slant height of the cone.

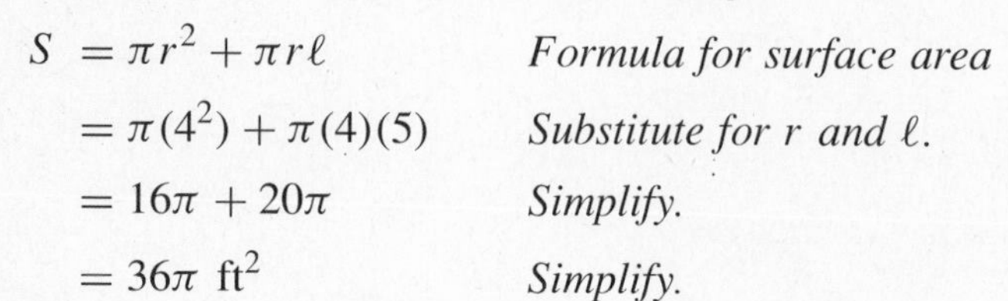

$S = \pi r^2 + \pi r \ell$ *Formula for surface area*

$= \pi(4^2) + \pi(4)(5)$ *Substitute for r and ℓ.*

$= 16\pi + 20\pi$ *Simplify.*

$= 36\pi \text{ ft}^2$ *Simplify.*

The surface area of the right cone is about 113.10 square feet.

- The lateral area of a pyramid or cone is the area of the lateral surface.
- The lateral area of a right cone is $\pi r \ell$.

EXERCISES

1. Find the surface area of the right cone shown below.

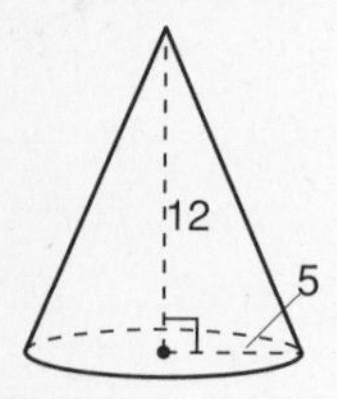

2. Find the lateral area of the right cone shown below.

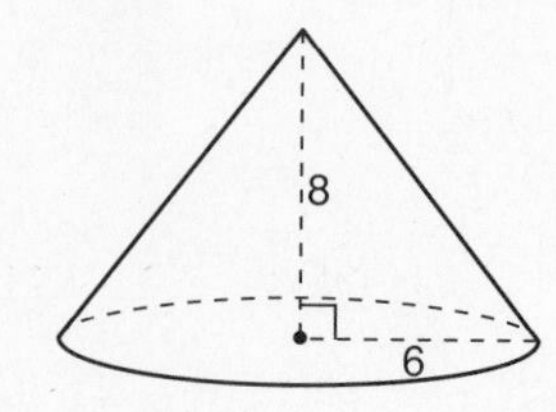

Reteach
Chapter 12

Name ______________________________

What you should learn:

12.4 How to use volume postulates to find the volume of a solid and find the volume of a prism and cylinder.

Correlation to Pupil's Textbook:

Chapter Test (p. 637)
Exercises 9, 10

Examples *Finding the Volume of a Solid, a Prism, and a Cylinder*

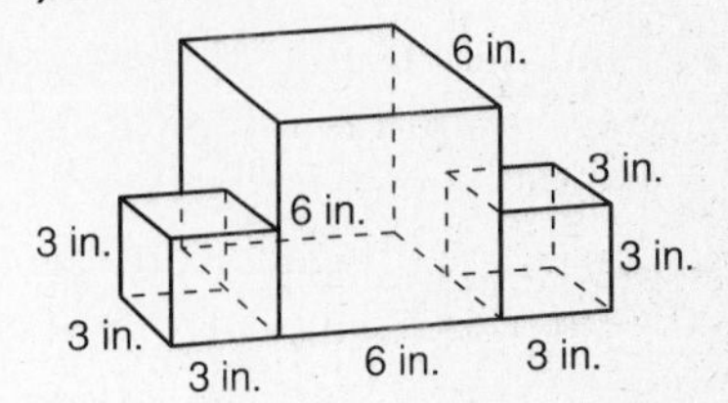

a. Find the volume of the solid shown at the right. State the volume postulates that you use in finding the volume of the solid.

1. Using the Volume of Cube Postulate, the volume of a cube is the cube of the length of its side, or $V = s^3$. The volume of the larger cube is $6^3 = 216$ cubic inches.
2. Using the Volume Congruence Postulate, two polyhedrons that are congruent have the same volume. The smaller cubes are congruent, therefore, they have the same volume. The volume of each smaller cube is $3^3 = 27$ cubic inches.
3. Using the Volume Addition Postulate, the volume of a solid is the sum of the volumes of all its nonoverlapping parts. The volume of the solid shown above is $216 + 2(27) = 270$ cubic inches.

b. Find the volume of the right prism and the right cylinder shown at the right.

1. The volume of a prism is $V = Bh$, where B is the area the base and h is the height. The base of the prism is an equilateral triangle. Its area, B, is $\frac{1}{4}s^2\sqrt{3} = \frac{1}{4}(6)^2\sqrt{3} = 9\sqrt{3}$ square inches. Then, $V = Bh = (9\sqrt{3})(4) = 36\sqrt{3}$ cubic inches ≈ 62.35 cubic inches.

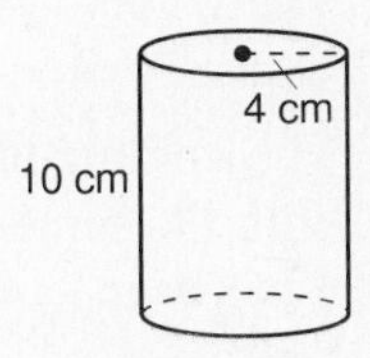

2. The volume of a cylinder is $V = Bh = \pi r^2 h$ where B is the area of the base, h is the height, and r is the radius of the base. The area of the base is $B = \pi r^2 = 16\pi$ square centimeters. So, the volume of the cylinder is $V = Bh = 16\pi(10) = 160\pi$ cubic centimeters ≈ 502.65 cubic centimeters.

Guidelines:

- Cavalieri's Principle states that if two solids have the same height and the same cross-sectional area at every level, then they have the same volume.

EXERCISES

1. Find the volume of a cube with 7-cm edges.

2. Find the volume of a prism with a triangular base with area of 32 ft^2 and a height of 3 feet.

3. Find the volume of a cylinder with a radius of 8 inches and height of 1 foot.

Reteach
Chapter 12

Name ______________________________

What you should learn:

12.5	How to find the volume of a pyramid and a cone and use volume to solve real-life problems

Correlation to Pupil's Textbook:

Chapter Test (p. 637)
Exercise 11

Examples *Finding the Volume of a Pyramid and a Cone*

a. Find the volume of the pyramid shown at the right. The base of the pyramid is a right triangle with legs of 6 m and 9 m. The height of the pyramid is 11 m.

The volume, V, of a pyramid is given by $V = \frac{1}{3}Bh$, where B is the area of the base and h is the height. The area of the base is 27 m^2, therefore, $V = \frac{1}{3}(27)(11) = 99$ cubic meters.

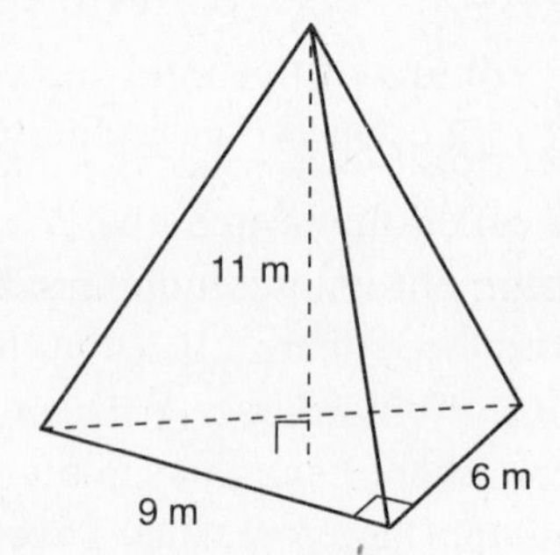

b. A highway construction pylon is a bright orange marker that is cone-shaped. Approximate the volume of a pylon that is 18 inches high and has a radius of 6 inches.

The volume, V, of a cone is given by $V = \frac{1}{3}Bh = \frac{1}{3}\pi r^2 h$, where B is the area of the base, h is the height, and r is the radius of the base. Then, $V = \frac{1}{3}\pi(6^2)(18) = 216\pi$ in.^{3}. The volume of the pylon is about 678.58 cubic inches.

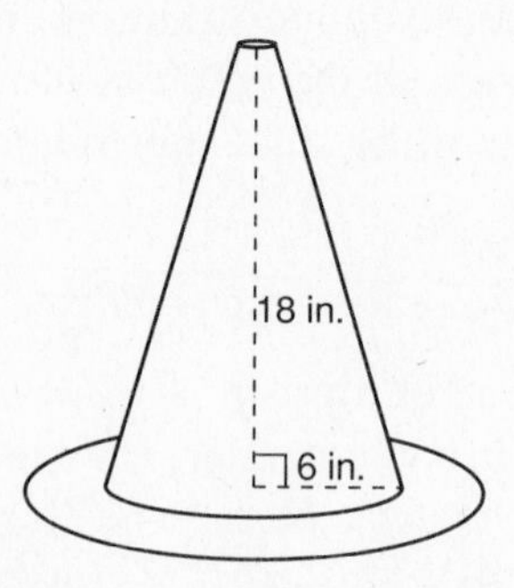

Guidelines:

- The volume of a pyramid is one-third of the volume of a prism with the same base area and the same height.
- The volume of a cone is one-third of the volume of a cylinder with the same base area and the same height.

EXERCISES

1. Find the volume of the oblique cone shown below. The radius of the base is 4.

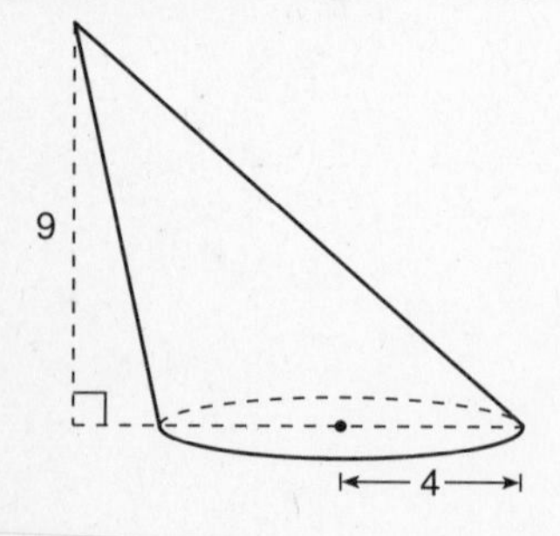

2. Find the volume of the pyramid with a rectangular base and a height of 10 shown below.

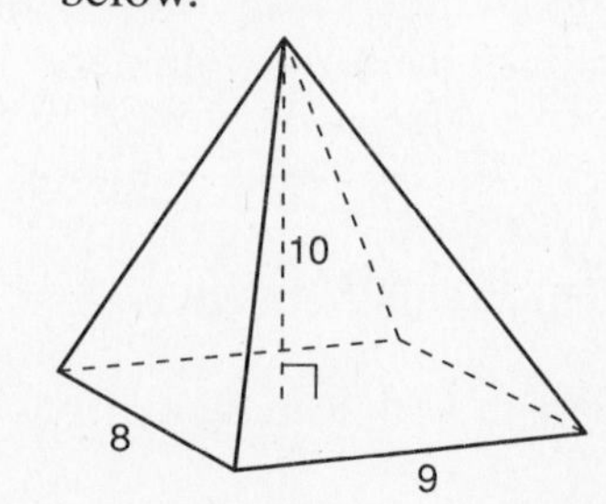

Reteach

Chapter 12

Name ______________________________

What you should learn:

12.6	How to find the surface area of a sphere and the volume of a sphere

Correlation to Pupil's Textbook:

Chapter Test (p. 637)

Exercises 7, 12, 14, 15, 17

Examples *Finding the Surface Area and Volume of a Sphere*

a. Match the following sphere definitions with the correct term.

1. the set of all points in space that are a given distance, r, from a point called the center — a. radius
2. any segment whose endpoints are the center of the sphere and a point on the sphere — b. hemispheres
3. a segment whose endpoints are on the sphere — c. great circle
4. a chord that contains the center of the sphere — d. diameter
5. the intersection of a plane and sphere such that the intersection contains the center of the sphere — e. chord
6. two congruent halves of a sphere formed by a great circle — f. sphere

Answers: 1. f 2. a 3. e 4. d 5. c 6. b

b. Find the surface area of a sphere with a radius of 3 centimeters.

The surface area, S, of a sphere of radius r is $S = 4\pi r^2$. Since the radius is 3 cm, $S = 4\pi(3)^2 = 36\pi$ cm^2. The surface area of the sphere is about 113.10 square centimeters.

c. Find the volume of a sphere with a radius of 6 inches.

The volume, V, of a sphere of radius r is $V = \frac{4}{3}\pi r^3$. Since the radius is 6 inches, $V = \frac{4}{3}\pi(6)^3 = 288\pi$ in.^{3}. The volume of the sphere is about 904.78 cubic inches.

Guidelines:

- All diameters of a sphere have the same length, d, which is twice the radius.
- If a plane intersects a sphere, the intersection will be either a circle or a single point.

EXERCISES

1. If the volume of a sphere is $\dfrac{500\pi}{3}$ in.^{3}, find its surface area.

2. If the surface area of a sphere is 196πm^3, find its volume.

3. If the circumference of the great circle of a sphere is 6π ft, find its volume.

4. If the volume of a sphere is $\dfrac{256\pi}{3}$ ft^3, find the volume of each hemisphere.

Reteach
Chapter 12

Name ____________________

What you should learn:

12.7	How to compare measures of similar solids and use similar solids to solve real-life problems

Correlation to Pupil's Textbook:

Chapter Test (p. 637)
Exercises 13, 16, 18, 19

Examples *Comparing Measures of Similar Solids and Using Similarity in Real Life*

a. The figure at the right shows two similar solids. Since they are similar, the ratios of corresponding linear measures are equal.

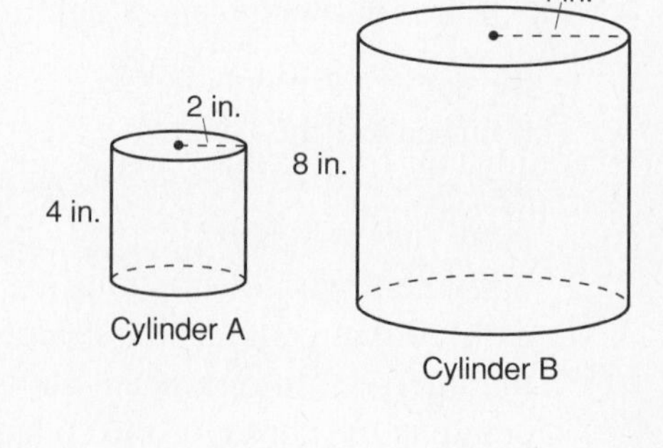

1. Find the scale factor of Cylinder A to Cylinder B.
   The scale factor is the ratio of corresponding radii or heights, 1 : 2.
2. Find the surface area and volume of Cylinder A.

$$S = 2\pi r^2 + 2\pi rh \qquad V = \pi r^2 h$$
$$= 2\pi(2^2) + 2\pi(2)(4) \qquad = \pi(2^2)(4)$$
$$= 8\pi + 16\pi \qquad = 16\pi \text{ in.}^3$$
$$= 24\pi \text{ in.}^2$$

3. Use the scale factor to find the surface area and volume of Cylinder B.
   If two solids are similar with a scale factor of 1 : 2, then their corresponding areas have a ratio of $1^2 : 2^2$ or 1 : 4 and their corresponding volumes have a ratio of $1^3 : 2^3$ or 1 : 8. The surface area of Cylinder B is $4(24\pi) = 96\pi$ square inches ≈ 301.59 square inches. The volume of Cylinder B is $8(16\pi) = 128\pi$ cubic inches ≈ 402.12 cubic inches.

b. You are making a cardboard megaphone, as shown at the right. You first make a cone with a height of 12 inches that has a base radius of 4 inches. You then cut a cone with a height of 2 inches and a base radius of $\frac{2}{3}$ inch from the top of the megaphone. Find the scale factor of the similar cones.

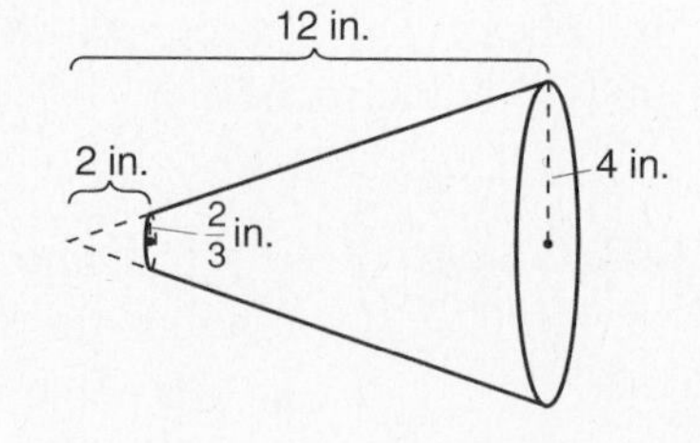

The scale factor is the ratio of corresponding radii or heights, 2 : 12 or $\frac{2}{3}$: 4. The scale factor is 1 : 6.

Guidelines:

- If two solids are similar with a scale factor of $a : b$, then corresponding areas have a ratio of $a^2 : b^2$ and corresponding volumes have a ratio of $a^3 : b^3$.

EXERCISE

Find the surface area and volume of a cube with 6-inch sides. Use a scale factor of 1 : 3 to find the surface area and volume of a larger similar cube.

Reteach

Chapter 13

Name ____________________

What you should learn:

13.1	How to find a locus in a plane and use a locus in constructions

Correlation to Pupil's Textbook:

Mid-Chapter Self-Test (p. 664) Exercises 1–4, 8–14

Chapter Test (p. 689) Exercises 1, 2

Examples *Finding a Locus in a Plane and Using a Locus in Constructions*

a. Locate two points X and Y in a plane. Describe the locus of all points in the plane that are equidistant from points X and Y.

As shown below, draw points X and Y. Locate several points that are equidistant from both X and Y. The set of all such points is a line that is the perpendicular bisector of $\overline{XY}$.

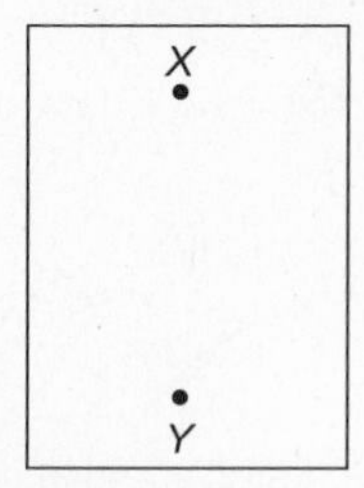

Draw given points.

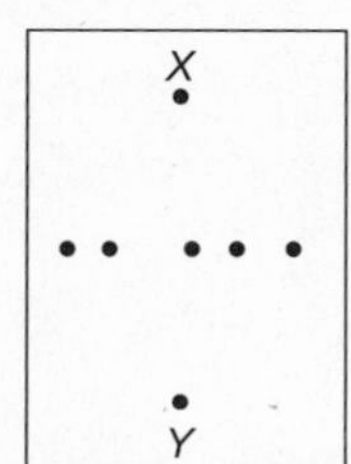

Locate several points.

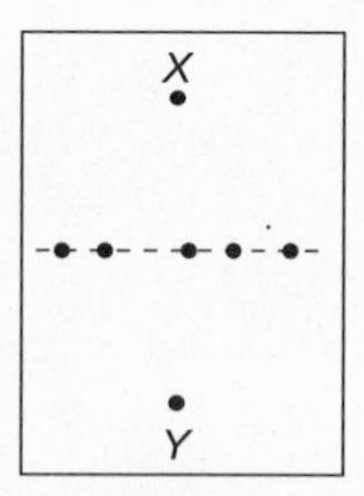

Recognize pattern.

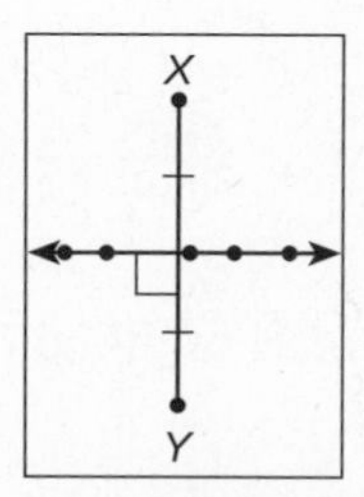

Draw the perpendicular bisector of $\overline{XY}$.

b. Sketch and describe the locus of the centers of all circles that are tangent to both sides of $\angle A$, as shown at the right.

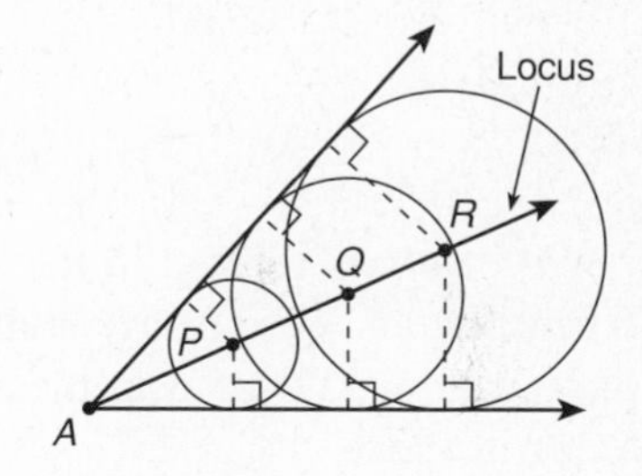

Locate P, Q, and R, the centers of three circles tangent to the sides of A. Since two radii of each circle are perpendicular to the sides of $\angle A$ at the points of tangency, centers P, Q, and R are equidistant from the sides of $\angle A$. Thus, the locus of the centers of circles tangent to both sides of $\angle A$ is $\overrightarrow{AR}$, the angle bisector of $\angle A$.

Guidelines:

- A *locus* in a plane is the set of all points in a plane that satisfy a given condition or a set of given conditions.
- A locus can be described as the path of an object that is moving in a plane according to some rule.

EXERCISES

In Exercises 1–4, sketch and describe the locus.

1. All points in a plane that are equidistant from the sides of a given triangle.
2. All points in a plane that are equidistant from two perpendicular lines ℓ and m.
3. All points in a plane that are 2 centimeters from point Q.
4. All points in a plane that are 1 inch from line k.

Reteach
Chapter 13

Name ____________________

What you should learn:

13.2 How to find a locus in space and use loci in real-life problems

Correlation to Pupil's Textbook:

Mid-Chapter Self-Test (p. 664)	**Chapter Test (p. 689)**
Exercises 5–7	Exercises 3, 4, 18

Examples *Finding a Locus in Space and Using Loci in Real Life*

a. Describe the locus of all points in space that are equidistant from the endpoints of $\overline{AB}$, as shown at the right.

In space, the locus is the plane that bisects $\overline{AB}$ and is perpendicular to $\overline{AB}$.

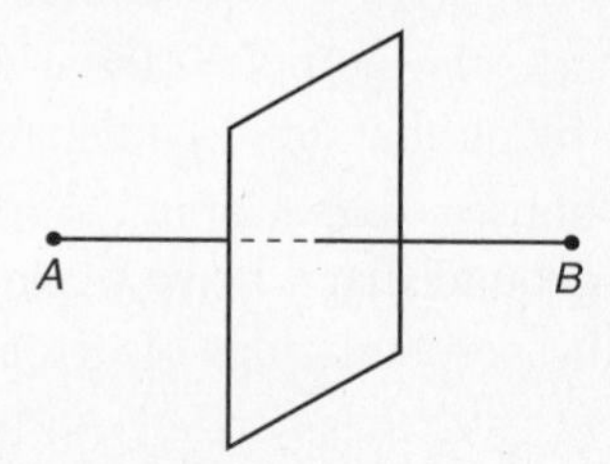

b. Describe the locus of all points in space that are 6 inches from a point, P.

In space, the locus is a sphere with a radius of 6 inches and center P.

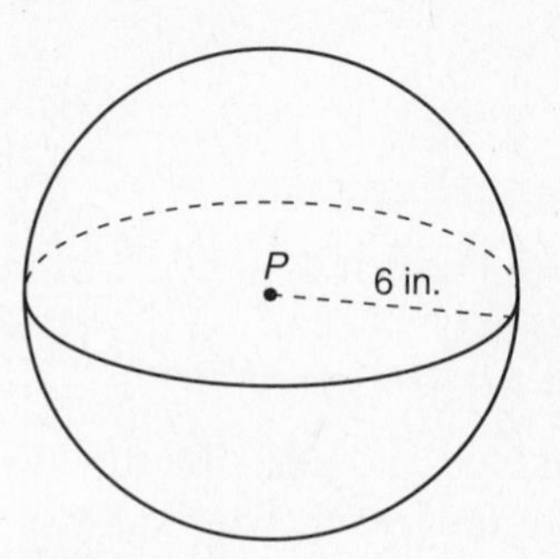

c. A caramel candy is formed in the shape of a sphere with a radius of 0.6 centimeters. The candy is dipped in chocolate so that the chocolate coating is 0.25 centimeters. Describe the chocolate coating as a locus in space.

The chocolate coating is the locus of all points in space that lie between 0.6 centimeters and 0.85 centimeters from the center of the candy.

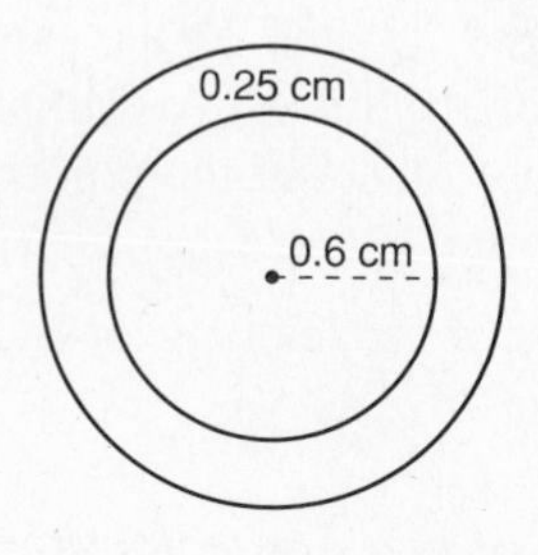

Guidelines:

- A locus of points "in a plane" is not the same as a locus of points "in space."

EXERCISES

In Exercises 1 and 2, describe the locus.

1. All points in space that are equidistant from the sides of a given triangle.

2. All points in space that are 2 centimeters from point Q.

Reteach

Chapter 13

Name ______________________________

What you should learn:

13.3	How to use loci to solve real-life problems and problems in coordinate geometry

Correlation to Pupil's Textbook:

Mid-Chapter Self-Test (p. 664)	**Chapter Test (p. 689)**
Exercises 8–10, 11–13, 15	Exercises 5, 6

Examples *Using Loci to Solve Problems in Real Life and in Coordinate Geometry*

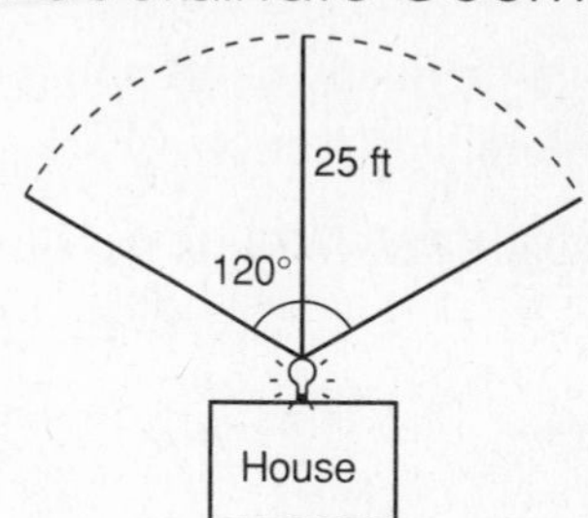

a. You are installing a security light at the back of your house, as shown in the sketch at the right. The light will project 25 feet from the house, in an arc of 120 degrees. Describe the locus of points on the ground that will be illuminated by the new light.

The locus of points is a sector of the circle, with a radius of 25 feet, centered at the light source. Since the locus is restricted by an arc of 120°, the locus is one-third the area of the circle.

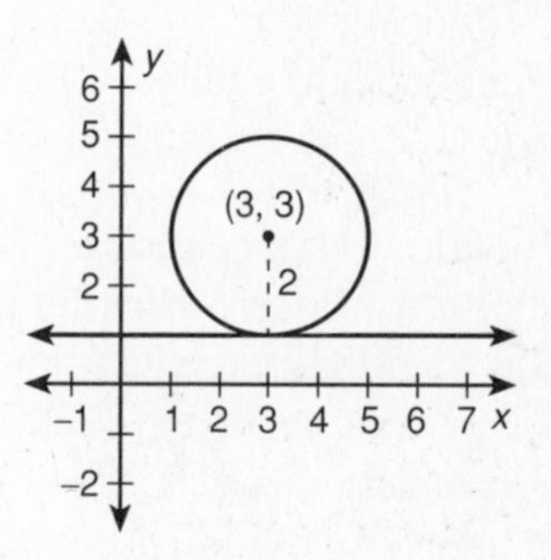

b. Determine the locus of points in a coordinate plane that are equidistant from the lines $y = -2$ and $y = 4$ and 2 units from the point (3, 3).

The compound locus is the locus of points that satisfy two conditions.

Condition 1: The locus of all points that are equidistant from the lines $y = -2$ and $y = 4$ is the line $y = 1$.

Condition 2: The locus of all points that are 2 units from the point (3, 3) is a circle, with the center of the circle at (3, 3) and a radius of 2. The intersection of the two loci is the point (3, 1).

Guidelines: When finding a compound locus,

- begin by sketching the locus corresponding to each condition.
- identify the point(s) of intersection of the individual loci.

EXERCISES

In Exercises 1–4, describe the compound locus in a coordinate plane.

1. All points 5 units from the origin and equidistant from the lines $y = 3$ and $y = 5$.

2. All points 5 units from the origin and equidistant from the lines $x = -5$ and $x = 5$.

3. All points 2 units from $P(2, 0)$ and 1 unit from $Q(5, 0)$.

4. All points equidistant from $P(0, 3)$ and $Q(3, 0)$ and 1 unit from the x-axis.

Reteach
Chapter 13

Name ______________________________

What you should learn:

13.4	How to use algebra to find loci and use loci to solve real-life problems

Correlation to Pupil's Textbook:

Chapter Test (p. 689)
Exercises 7–10, 15–17

Examples *Using Algebra to Find Loci and Using Loci in Real Life*

a. Find the locus of all points that lie on the graphs of both $2x - 3y = -7$ and $3x + y = -5$.

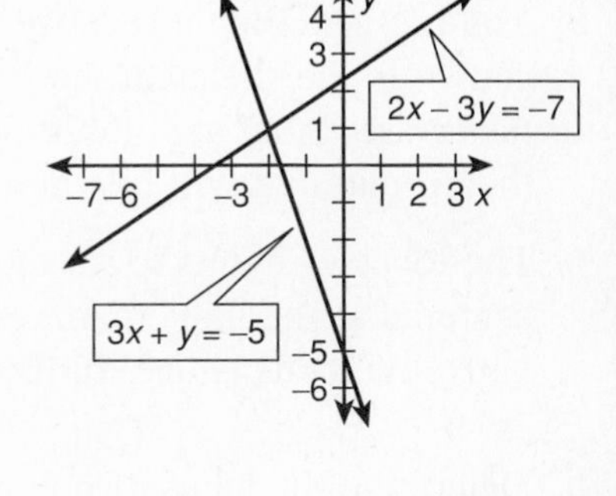

Solve the system of equations using linear combinations.

$$\begin{aligned} 2x - 3y = -7 &\Rightarrow 2x - 3y = -7 && \text{Equation 1} \\ 3x + y = -5 &\Rightarrow 9x + 3y = -15 && \text{Multiply Equation 2 by 3.} \\ & \quad 11x = -22 && \text{Add equations.} \end{aligned}$$

It follows that $x = -2$. By substituting $x = -2$ into either equation, you conclude that $y = 1$. The system has one solution, the ordered pair $(-2, 1)$. So the locus is the point $(-2, 1)$.

b. A parking garage charges \$3.00 for the first hour and \$1.00 for each additional hour. A nearby parking lot charges \$2.00 for the first hour and \$1.50 for each additional hour. Write mathematical models for the parking charges. Sketch the graphs of the models on the same coordinate plane. When are the charges equal? If x represents the number of additional hours, the mathematical models are

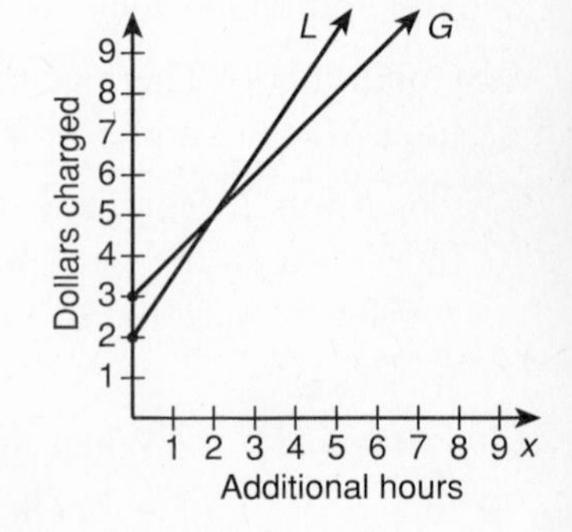

$G = 3.00 + 1.00x$ *Parking garage charges*

$L = 2.00 + 1.50x$ *Parking lot charges*

The charges are equal when the equations have the same solution, at the ordered pair $(2, 5)$.

Answer: For a 3-hour stay, you will pay the same amount, \$5.00, at either parking facility.

Guidelines:

- The locus of points that satisfy two linear equations is a point (if the lines intersect).
- The locus of points in a coordinate plane that are equidistant from two points is a line.
- The locus of points that satisfy three linear inequalities in a coordinate plane is the intersection of three half-planes.

EXERCISES

1. Find the locus of all points in a coordinate plane that are equidistant from $C(0, -2)$ and $D(2, 0)$.

2. Find the locus of points that satisfy all three inequalities.

$$\begin{cases} x - y < 2 \\ x > -2 \\ y \leq 3 \end{cases}$$

Reteach
Chapter 13

Name ______________________________

What you should learn:

13.5	How to find cross sections of a solid and to use cross sections to solve real-life problems

Correlation to Pupil's Textbook:

Chapter Test (p. 689)
Exercises 11–14

Examples *Finding Cross Sections of a Solid and Using Cross Sections in Real Life*

a. Sketch the following cross sections of a right triangular prism, as shown at the right.

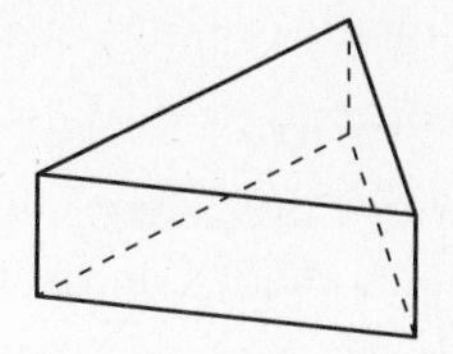

a. a triangle b. a rectangle c. a line segment

1. 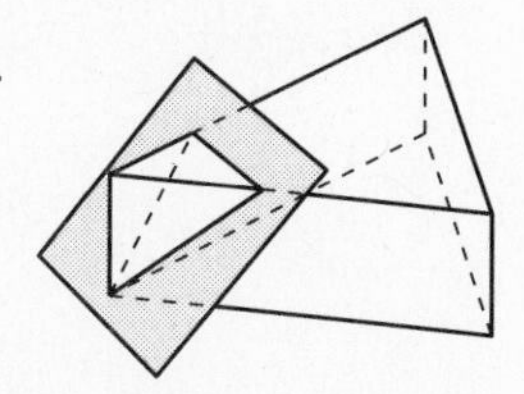

a triangle

2. 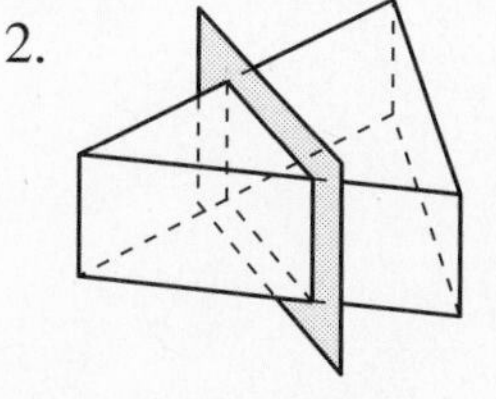

a rectangle

3. 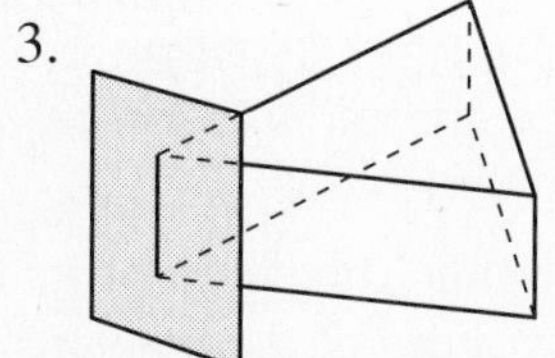

a line segment

b. Conic sections are cross sections of a double napped right cone. Match the following conic sections with its cross section.

1. 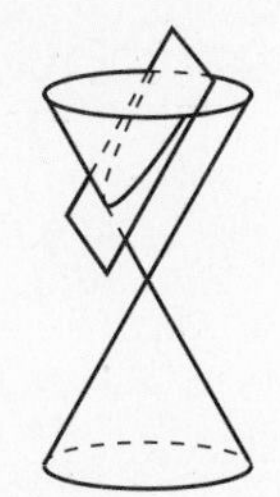

2. 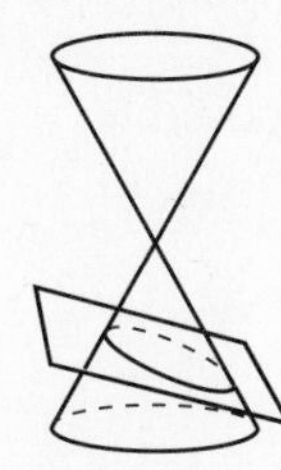

3. 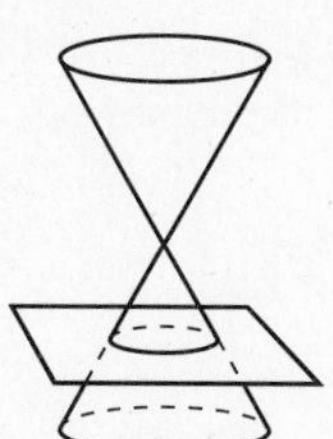

4. 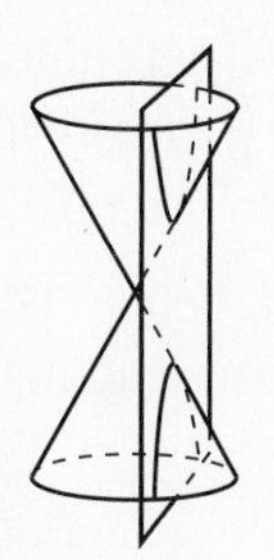

a. ellipse
b. hyperbola
c. parabola
d. circle

Answers: 1. c 2. a 3. d 4. b

c. A skating-rink, shown at the right, is built in the shape of an ellipse. If F_1 and F_2 are the foci and $d_1 + d_2 = 42$ meters, find $d_3 + d_4$.

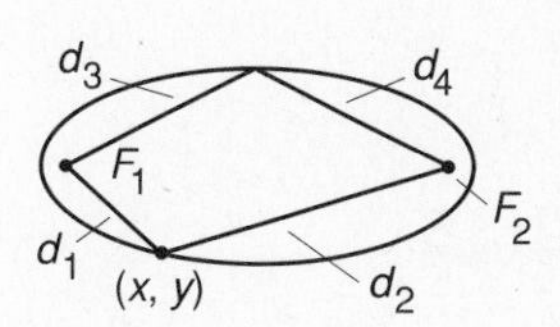

By definition, an ellipse is the locus of points (x, y) such that the sum of the distances between (x, y) and two distinct foci, F_1 and F_2, is constant. Since $d_1 + d_2 = 42$ meters, $d_3 + d_4 = d_1 + d_2 = 42$ meters.

Guidelines:

- A cross section is the intersection of a plane and a solid.
- The four basic conic sections are a circle, an ellipse, a parabola, and a hyperbola.
- Degenerate conic sections are a point, a line, and a pair of intersecting lines.

EXERCISE

Sketch the possible cross sections of a cube.

Reteach

Chapter 13

Name ______________________________

What you should learn:

13.6	How to become a better problem solver by reviewing concepts and making connections

Correlation to Pupil's Textbook:

Chapter Test (p. 689)
Exercises 15–17

Examples *Reviewing Important Concepts and Making Connections*

a. Geometry is

- a problem-solving language used to answer questions about real life
- a logical system with rules that can be applied to all branches of mathematics and science
- a system of measures
- active exploration and conjectures based on observations

b. Geometry is connected to other branches of mathematics. You use geometry in the following areas of mathematics—

- coordinate geometry which uses geometry and algebra
- trigonometry which uses circles, angles, and right triangles
- calculus which uses slopes and areas in differentiation and integration

c. There are many famous geometry problems that have not been solved. Three famous construction problems that have been shown to be impossible are—

- the trisection of an angle by two constructed rays
- the doubling of a cube by constructing the side of a cube whose volume is twice that of a given cube
- the squaring of a circle by constructing the side of a square whose area equals that of a given circle

EXERCISE

Write a short paragraph about the important concepts of geometry which are relevant in your life. Discuss what you have learned about logical reasoning and problem-solving skills.

Answers to Exercises

Lesson 1.1 (page 1)

1. The lower curve of figure **A** looks smaller because the viewer's eye compares it to the longer upper curve of figure **B**.
2. Triangle; bars meet at right angles; yes, a triangle with 3 right angles does not exist.
3. A, H, a combination of A and E, and a combination of C and G

Lesson 1.2 (page 2)

1. $-13, 1, -3, -5, 9$ 2. 16, 1, 10, 9

Lesson 1.3 (page 3)

1. Answers will vary.

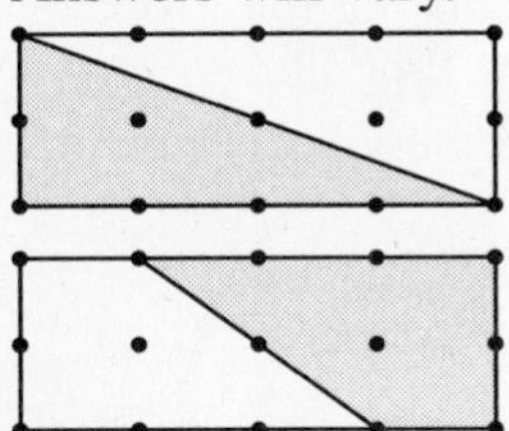

2. a and c 3. a and b

Lesson 1.4 (page 4)

1. Rotational symmetry of 45°, 90°, 135°, and 180°
2. Vertical line of symmetry
3. Horizontal line of symmetry

Lesson 1.4 (page 5)

1. Midpoint of $\overline{AB} = \left(2, \frac{3}{2}\right)$
   Midpoint of $\overline{BC} = \left(2, -\frac{3}{2}\right)$
   Midpoint of $\overline{CD} = \left(-2, -\frac{3}{2}\right)$
   Midpoint of $\overline{DA} = \left(-2, \frac{3}{2}\right)$
2. 14

Lesson 1.5 (page 6)

1. $m = -1$ 2. $m = 1$
3. $m = -1$
4. a. perpendicular
   b. neither
   c. parallel

Lesson 1.5 (page 7)

1. Answers will vary. 2. Answers will vary.

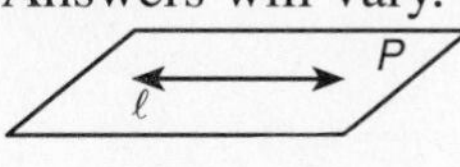

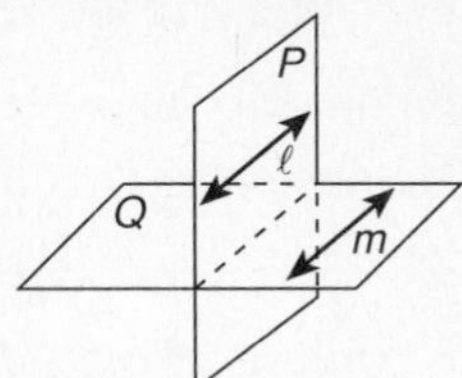

3. a. perpendicular
   b. coplanar
   c. parallel

Lesson 1.6 (page 8)

1. Area of floor = Length of room · Width of room
2. Area of floor = A (square feet)
   Length of room = 11.5 (feet)
   Width of room = 10.5 (feet)
3. $A = l \cdot w$ 4. $A = (11.5)(10.5)$
   $= 120.75$
5. You need 120.75 square feet of linoleum.

Lesson 1.7 (page 9)

1.

2.

Lesson 2.1 (page 10)

1. $\overrightarrow{MN}$ and $\overrightarrow{ML}$ 2. $\angle 1$ 3. $\angle 2$
4. $\angle 1$ and $\angle 2$ 5. $\angle LMN$
6. 12, 15, 18; $3n$

Lesson 2.2 (page 11)

1. 2 2. 2 3. 8
4. $m\angle TSV = 40°$

Lesson 2.3 (page 12)

1. 5 **2.** $(-\frac{1}{2}, 1)$ **3.** Always

4. Always **5.** Sometimes

Lesson 2.4 (page 13)

1. Hypothesis: $m\angle C = 100°$
Conclusion: $\angle C$ is obtuse.
Converse: If $\angle C$ is obtuse, then $m\angle C = 100°$.

2. Hypothesis: Points K and L are different points in a plane.
Conclusion: There is a third point in a plane not on $\overleftrightarrow{KL}$.
Converse: If there is a third point in a plane not on line $\overleftrightarrow{KL}$, then points K and L are different points in a plane.

3. Hypothesis: Two distinct planes intersect.
Conclusion: Their intersection is a line.
Converse: If the intersection of two planes is a line, then the planes are distinct.

4. Hypothesis: Points P, Q, and R are noncollinear.
Conclusion: They lie in one and only one plane.
Converse: If points P, Q, and R lie in one and only one plane, then they are noncollinear.

5. Two lines are perpendicular if and only if they intersect to form right angles.

Lesson 2.5 (page 14)

1. b **2.** d

3. a or e **4.** a or e **5.** c

6. d **7.** a **8.** b **9.** c

Lesson 2.6 (page 15)

1.

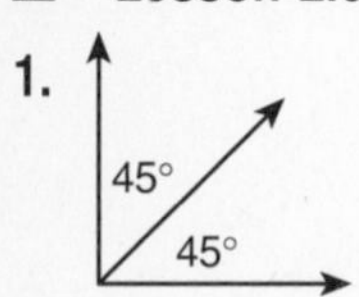

2.

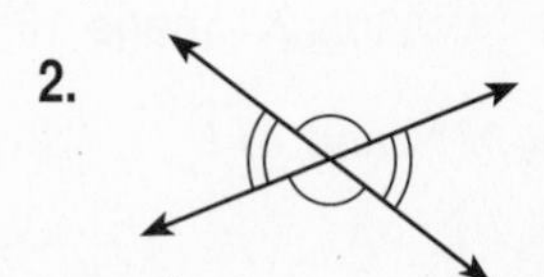

3. Answers will vary.

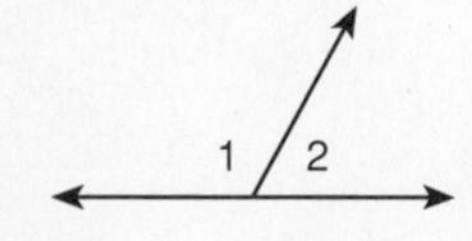

4. Answers will vary.

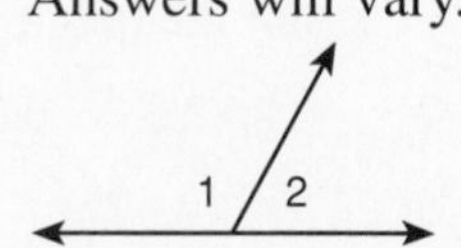

5.

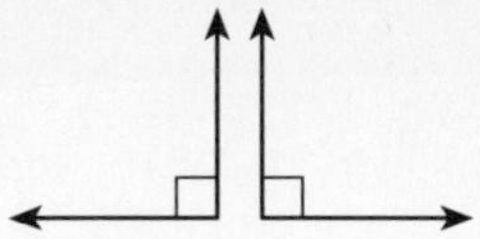

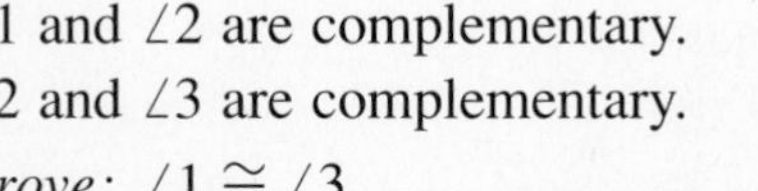

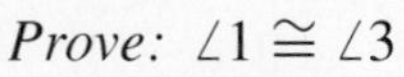

6. Not possible

Lesson 2.6 (page 16)

1. *Given:*
$\angle 1$ and $\angle 2$ are complementary.
$\angle 2$ and $\angle 3$ are complementary.

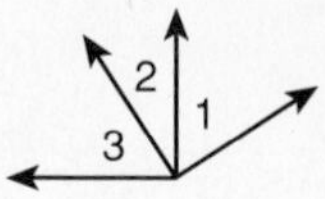

Prove: $\angle 1 \cong \angle 3$

Statements	*Reasons*
1. $\angle 1$ and $\angle 2$ are complementary.	**1.** Given
2. $m\angle 1 + m\angle 2 = 90°$	**2.** Def. of complementary ∠s
3. $\angle 2$ and $\angle 3$ are complementary.	**3.** Given
4. $m\angle 2 + m\angle 3 = 90°$	**4.** Def. of complementary ∠s
5. $m\angle 1 + m\angle 2 = m\angle 2 + m\angle 3$	**5.** Substitution Prop. of Equality
6. $m\angle 1 = m\angle 3$	**6.** Subtraction Prop. of =
7. $\angle 1 \cong \angle 3$	**7.** Def. of Congruent ∠s

2. *Given:*
$\angle 1$ and $\angle 2$ are complementary.
$\angle 3$ and $\angle 4$ are complementary.
$\angle 3 \cong \angle 2$

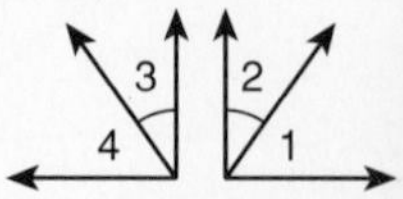

Prove: $\angle 1 \cong \angle 4$

Statements	*Reasons*
1. $\angle 1$ and $\angle 2$ are complementary.	**1.** Given
2. $m\angle 1 + m\angle 2 = 90°$	**2.** Def. of complementary ∠s
3. $\angle 3$ and $\angle 4$ are complementary.	**3.** Given
4. $m\angle 3 + m\angle 4 = 90°$	**4.** Def. of complementary ∠s
5. $m\angle 1 + m\angle 2 = m\angle 3 + m\angle 4$	**5.** Substitution Prop. of Equality
6. $\angle 3 \cong \angle 2$	**6.** Given
7. $m\angle 3 = m\angle 2$	**7.** Def. of congruence
8. $m\angle 1 = m\angle 4$	**8.** Subtraction Prop. of =
9. $\angle 1 \cong \angle 4$	**9.** Def. of congruence

Lesson 3.1 (page 17)

1. $\overleftrightarrow{MO} \parallel \overleftrightarrow{RT}$ 2. $\overleftrightarrow{MO} \parallel \overleftrightarrow{NP}$
3. Parallel or nonintersecting
4. Intersecting 5. Skew lines

Lesson 3.2 (page 18)

1. There are many solutions. The lines are coincident.
2. The lines are parallel

Lesson 3.2 (page 19)

1. $y = -\frac{2}{5}x + \frac{11}{5}$
2. $y = -\frac{1}{2}x + 8$

Lesson 3.3 (page 20)

1. $\angle 1$ is not a rt. angle.
2. The measure of $\angle 1$ is not 90°.
3. If Jared does not buy a compact disc player, then Jared did not earn $150.
4. If the sailing race is not cancelled, then the weather is not too windy.

Lesson 3.3 (page 21)

1. Law of Detachment
2. Law of Syllogism
3. We will go skiing. Law of Detachment
4. We will need motel reservations. Law of Syllogism

Lesson 3.4 (pages 22 and 23)

1. *Given:* $m\angle 1 = m\angle 3$
Prove: $m\angle QPS = m\angle TPR$
   - If $m\angle 1 = m\angle 3$, then $m\angle 1 + m\angle 2 = m\angle 2 + m\angle 3$ by Addition Prop. of Equality.
   - By the Angle Addition Postulate, $m\angle 1 + m\angle 2 = m\angle QPS$ and $m\angle 2 + m\angle 3 = m\angle TPR$.
   - Using the Subsitution Prop. of Equality, $m\angle QPS = m\angle TPR$.

2. *Given:* $\angle 1$ and $\angle 2$ are right angles.
Prove: $\angle 1 \cong \angle 2$
Proof:

Statements	*Reasons*
1. $\angle 1$, $\angle 2$ are rt. $\measuredangle$s.	1. Given
2. $m\angle 1 = 90^\circ$	2. Def. of rt. angle
3. $m\angle 2 = 90^\circ$	3. Def. of rt. angle
4. $m\angle 1 = m\angle 2$	4. Substitution Prop. of Equality
5. $\angle 1 \cong \angle 2$	5. Def. of $\cong$ angles

3. 1. c or d 2. a or f 3. c or d
4. a or f 5. b 6. g 7. e

Lesson 3.4 (page 24)

Luke and Juan drove trucks. Dedra, Amber, and Carlo drove automobiles.

Lesson 3.5 (page 25)

1. **a.** 65° **b.** 115° **c.** 65° **d.** 115°
e. 115° **f.** 65° **g.** 115°

Lesson 3.6 (page 26)

1. If two parallel lines are cut by a transversal, then consecutive interior angles are supplementary. Yes, the converse is true.
2. If two parallel lines are cut by a transversal, then alternate exterior angles are congruent. Yes, the converse is true.

Lesson 3.7 (page 27)

1. $\vec{v} + \vec{w} = \langle 0, 5\rangle$
2. $\vec{u} + \vec{v} = \langle 0, 0\rangle$
3. $\vec{z} + \vec{y} = \langle 5, 0\rangle$
4. $\vec{s} + \vec{t} = \langle 5, 5\rangle$
5. $\overrightarrow{AB} = \langle 3, 0\rangle, \overrightarrow{CD} = \langle -3, 0\rangle,$
$AB = 3, CD = 3, \neq$
6. $\overrightarrow{AB} = \langle 4, -3\rangle, \overrightarrow{CD} = \langle 4, -3\rangle,$
$AB = 5, CD = 5, =$

Lesson 3.7 (page 28)

1. $\vec{v} \cdot \vec{w} = -4 + 4 = 0, \perp$
2. $\vec{u} \cdot \vec{v} = -9 - 36 = -45, \not\perp$
3. $\vec{z} \cdot \vec{y} = 0 - 16 = -16, \not\perp$
4. $\vec{s} \cdot \vec{t} = -6 + 6 = 0, \perp$

■ Lesson 4.1 (page 29)

1. $m\angle R$ **2.** $\overline{QR}$ **3.** $\overline{EF}$

4. $\overline{EG}$ **5.** $\overline{FH}$ and $\overline{GH}$ **6.** $\triangle EFH$

■ Lesson 4.2 (page 30)

1. 50° **2.** 50° **3.** 90°

4. 40° **5.** 90° **6.** 40°

7.

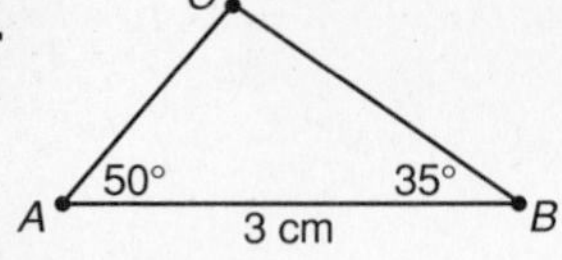

8.

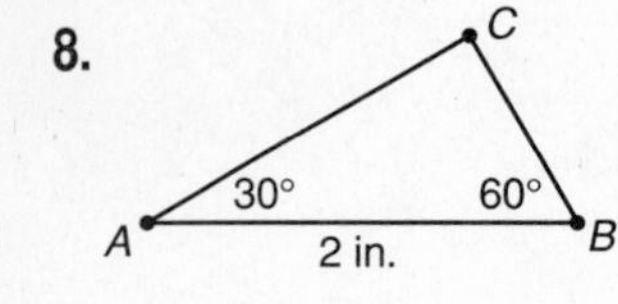

■ Lesson 4.3 (pages 31 and 32)

1.

Statements	*Reasons*
1. $\overline{FG}$ and $\overline{JK}$ bisect each other at P.	**1.** Given
2. $\overline{FP} \cong \overline{GP}$ $\overline{KP} \cong \overline{JP}$	**2.** Def. of bisect
3. $\angle 1 \cong \angle 2$	**3.** Vertical ∠s are ≅.
4. $\triangle FKP \cong \triangle GJP$	**4.** SAS Cong. Post.

2.

Statements	*Reasons*
1. $\overline{FJ} \cong \overline{KG}$	**1.** Given
2. $\overline{FJ} \parallel \overline{KG}$	**2.** Given
3. $\angle 3 \cong \angle 4$	**3.** If lines are ∥, then alternate interior ∠s are ≅.
4. $\overline{FG} \cong \overline{FG}$	**4.** Reflexive Prop. of Congruence
5. $\triangle FJG \cong \triangle GKF$	**5.** SAS Cong. Post.

3. 1. b, f, or i 6. a
2. e 7. b, f, or i
3. d or h 8. b, f, or i
4. d or h 9. c
5. g

■ Lesson 4.4 (pages 33 and 34)

1. ASA **2.** AAS **3.** AAS **4.** ASA

5. $\overline{BC} \cong \overline{EF}$ or $\overline{AC} \cong \overline{DF}$

6. $\overline{AB} \cong \overline{DE}$

■ Lesson 4.5 (pages 35 and 36)

1. *Planning a proof:* Mark the diagram with the given information $\overline{PQ} \parallel \overline{RS}$ and $\overline{PQ} \cong \overline{RS}$. Since $\overline{PQ} \parallel \overline{RS}$, alternate interior angles $\angle 1$ and $\angle 2$ are congruent. $\overline{PR} \cong \overline{PR}$ by Reflexive Prop. of Congruence. Since is is given that $\overline{PQ} \cong \overline{RS}$, $\triangle PQR \cong \triangle RSP$ by the SAS Congruence Postulate. By CPCTC, $\angle 3 \cong \angle 4$. Finally, since alternate interior angles are ≅, we conclude $\overline{PS} \parallel \overline{QR}$.

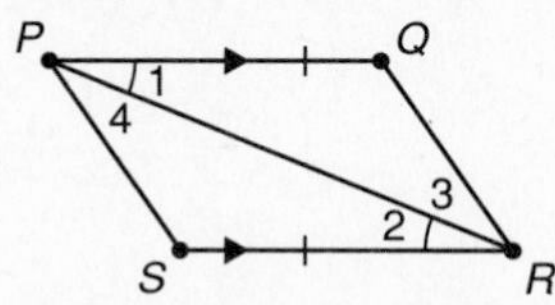

2. 1. e or g 2. a or f 3. e or g
4. a or f 5. d 6. c 7. b

■ Lesson 4.6 (page 37)

Given: $\overline{AB} \cong \overline{BC} \cong \overline{CA}$

Prove: $\angle A \cong \angle B \cong \angle C$

Statements	*Reasons*
1. $\overline{BC} \cong \overline{CA}$	**1.** Given
2. $\angle A \cong \angle B$	**2.** Base Angles Thm.
3. $\overline{CA} \cong \overline{AB}$	**3.** Given
4. $\angle B \cong \angle C$	**4.** Base Angles Thm.
5. $\angle A \cong \angle C$	**5.** Transitive Property of Congruence
6. $\triangle ABC$ is equiangular.	**6.** If $\angle A \cong \angle B \cong \angle C$, then △ is equiangular by definition.

■ Lesson 4.7 (page 38)

1. Given
2. Given
3. Given
4. SSS Congruence Postulate
5. CPCTC
6. Corresponding ∠s Converse Post.

Lesson 5.1 (pages 39 and 40)

1. $\angle XPQ$ and $\angle YPQ$ are right angles.
2. $\overline{RX} \cong \overline{RY}$ 3. $\overline{XP} \cong \overline{YP}$
4. $\angle XRP \cong \angle YRP$ 5. $y = -4x + 33$
6. $XQ = YQ$ 7. Yes
8. Slope of $\overrightarrow{XW} = \frac{2}{3}$
   Slope of $\overline{WZ} = -\frac{3}{2}$
   $\frac{2}{3} = -\frac{1}{-\frac{3}{2}}$
9. Slope of $\overrightarrow{XY} = -\frac{2}{3}$
   Slope of $\overline{YZ} = \frac{3}{2}$
   $-\frac{2}{3} = -\frac{1}{\frac{3}{2}}$
10. $\sqrt{29.25}$ 11. $\sqrt{29.25}$
12. Z lies in the interior of $\angle WXY$ and is equidistant from the sides of the angles.

Lesson 5.2 (pages 41 and 42)

1. Circumcenter 2. Incenter
3. Centroid
4. Centroid
5. 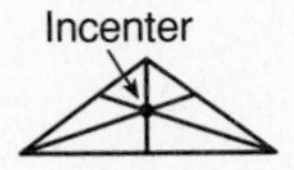

6. 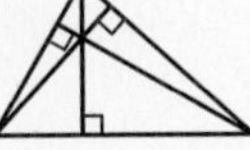
7. 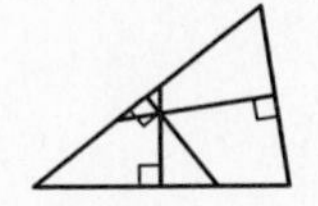

Lesson 5.3 (page 43)

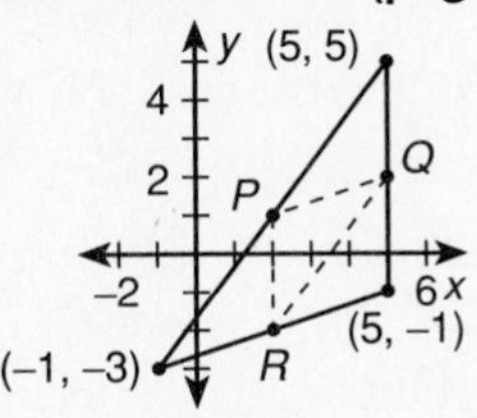

The coordinates of the vertices of the triangle are $(5, 5)$, $(5, -1)$ and $(-1, -3)$.

Lesson 5.4 (page 44)

1. 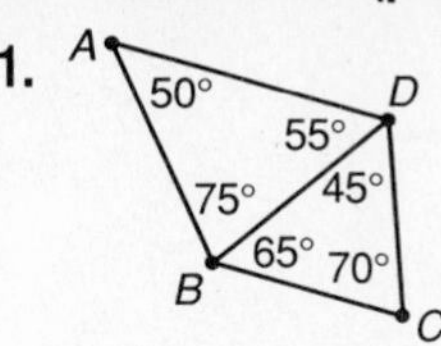

$\overline{BC}, \overline{CD}, \overline{BD}, \overline{AB}, \overline{AD}$

2. $\angle X, \angle W, \angle Y$

Lesson 5.5 (pages 45 and 46)

1. If $\overline{AB} \cong \overline{XY}$, $\overline{BC} \cong \overline{XZ}$, $AC > YZ$, then $m\angle B > m\angle X$. Answers will vary.

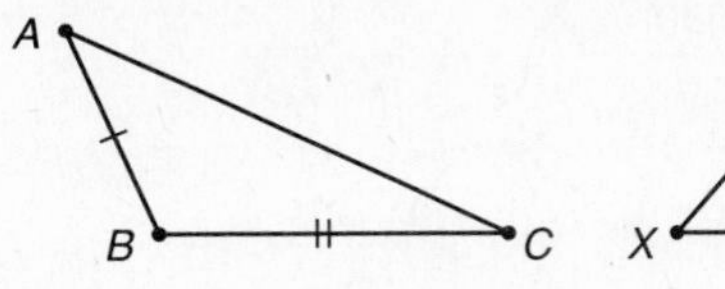

2. $AC > BD$ 3. $ED > BC$
4. $AB > BE$
5. Suppose that ℓ is perpendicular to m. Then angles $\angle 1$ and $\angle 2$ are right angles which are congruent. This contradicts the given statement that $\angle 1 \ncong \angle 2$. The case that ℓ is perpendicular to m is not true and the case the ℓ is not perpendicular to m is true.

Lesson 5.6 (page 47)

1. 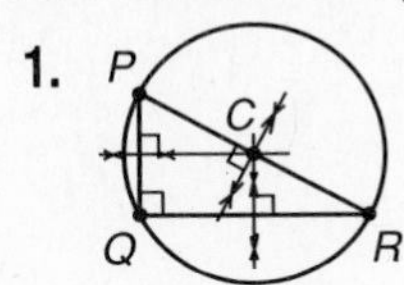

Circumscribed circle

2. You can construct an isosceles triangle $\triangle XYZ$. The altitude $\overline{YW}$ is also a perpendicular bisector of $\overline{XZ}$.

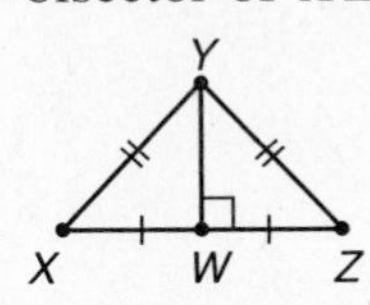

Lesson 6.1 (page 48)

1. Regular
   Vertices: V, W, X, Y, Z
   Diagonals: $\overline{VY}, \overline{VX}, \overline{WY}, \overline{WZ}, \overline{XZ}$
2. Equilateral
   Vertices: C, D, E, F, G, H
   Diagonals: $\overline{CE}, \overline{CF}, \overline{CG}, \overline{DH}, \overline{DG}, \overline{DF}, \overline{EH}, \overline{EG}, \overline{FH}$
3. Regular
   Vertices: P, Q, R, S
   Diagonals: $\overline{PR}, \overline{SQ}$
4. Equiangular
   Vertices: J, K, L, M, N
   Diagonals: $\overline{JM}, \overline{JL}, \overline{KM}, \overline{KN}, \overline{LN}$

Lesson 6.2 (pages 49 and 50)

1. 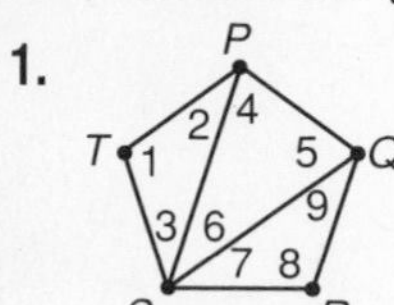

$m\angle 1 = 108°$ $m\angle 6 = 36°$
$m\angle 2 = 36°$ $m\angle 7 = 36°$
$m\angle 3 = 36°$ $m\angle 8 = 108°$
$m\angle 4 = 72°$ $m\angle 9 = 36°$
$m\angle 5 = 72°$

2. 1080° **3.** 135° **4.** 360° **5.** 45°
6. 75° **7.** 120° **8.** 40°

Lesson 6.3 (pages 51 and 52

1. 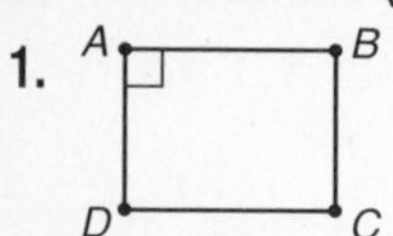

Conclusion–All the angles are right angles. If $m\angle A = 90°$, then $m\angle C = 90°$. Since $m\angle A + m\angle D = 180°$, $m\angle D = 90°$. If $m\angle D = 90°$, then $m\angle B = 90°$.

2. 6 **3.** 14 **4.** 10° **5.** 34°
6. 44° **7.** 132° **8.** 48°
9. $\angle GAB \cong \angle DEF$
   Since opposite angles of $\square ABCG$ are congruent, $\angle GAB \cong \angle BCG$. But vertical angles $\angle BCG$ and $\angle DCF$ are congruent. Therefore, $\angle GAB \cong \angle DCF$ by transitivity. In $\square CDEF$, opposite angles $\angle DCF \cong \angle DEF$. By transitivity, $\angle GAB \cong \angle DEF$.

Lesson 6.4 (pages 53 and 54)

1. Show that $\overline{HG} \cong \overline{EF}$.
   $HG = \sqrt{(4-7)^2 + (4-2)^2} = \sqrt{13}$
   $EF = \sqrt{(-1-2)^2 + (1+1)^2} = \sqrt{13}$
   From Example b, $\overline{EH} \cong \overline{FG}$.
   Since $\overline{HG} \cong \overline{EF}$ and $\overline{EH} \cong \overline{FG}$, by Theorem 6.7, $EFGH$ is a parallelogram.
2. 1. f or g 2. c
   3. b 4. f or g
   5. h 6. a or d
   7. a or d 8. e

Lesson 6.5 (page 55)

$12 \times 9 = 108$ inches or 9 feet

Lesson 6.6 (pages 56 and 57)

1. 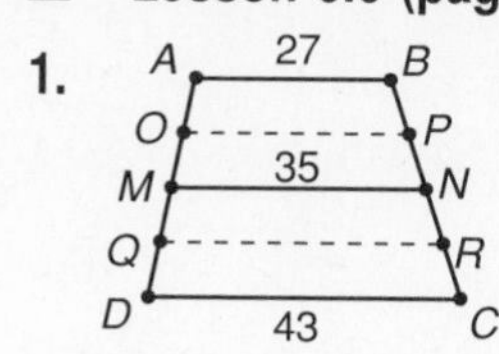

$OP = \frac{1}{2}(27 + 35) = 31$
$QR = \frac{1}{2}(35 + 43) = 39$

2. $m\angle Z = 115°$
   $m\angle Y = 115°$
3. $ON = 14$ Length of midsegment is 9.
4. 1. b or d 2. b or d
   3. a 4. e
   5. f 6. c

Lesson 6.7 (page 58)

1. From Example a on page 58, you know that all sides of both rhombuses are congruent and you know that $\angle N \cong \angle X$. You would show that opposite angles $\angle M \cong \angle O$ and $\angle W \cong \angle Y$. Since it is given that $\angle M \cong \angle W$, you can show that $\angle O \cong \angle Y$. Now you have three angles and included sides of one quadrilateral congruent to the corresponding three angles and included sides of another quadrilateral. You can use the ASASA Congruence Theorem.
2. A kite cannot have two pairs of opposite congruent angles, only one pair.

Lesson 7.1 (pages 59 and 60)

1. isometry, reflection
2. not an isometry
3. isometry, translation

Lesson 7.2 (pages 61 and 62)

1.

2. 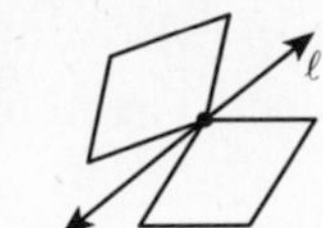

3. 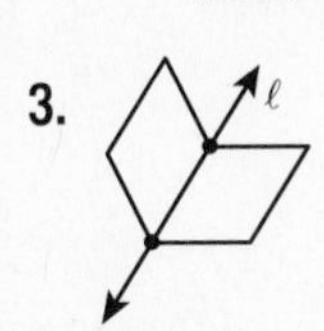

Lesson 7.3 (pages 63 and 64)

1. 27°

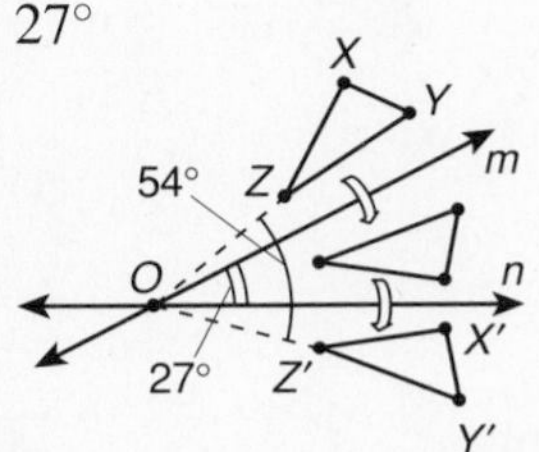

2. 108°

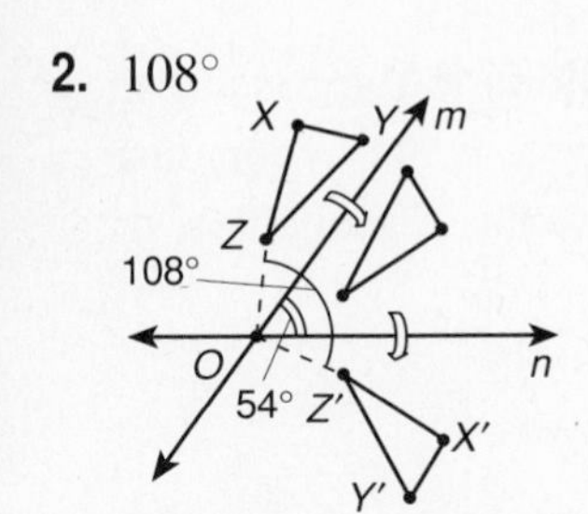

3. 60°, 120°, or 180° clockwise or counterclockwise
4. 72° or 144° clockwise or counterclockwise
5. 90° or 180° clockwise or counterclockwise
6. 120° clockwise or counterclockwise

Lesson 7.4 (pages 65 and 66)

1. b 2. a 3. c 4. 6

5. $\overline{AA'}$ or $\overline{BB'}$ or $\overline{AA''}$ or $\overline{BB''}$ or $\overline{A'A''}$ or $\overline{B'B''}$

Lesson 7.5 (pages 67 and 68)

1. 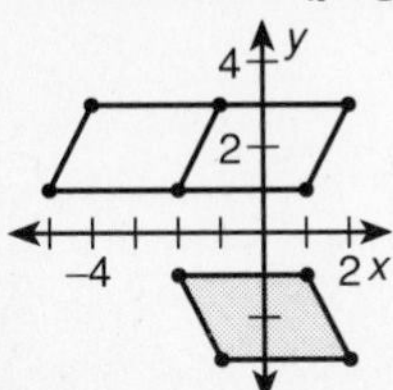

2. 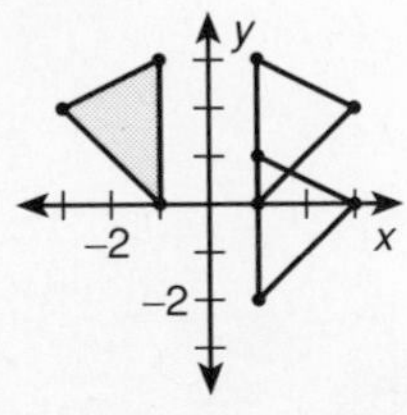

3.

Lesson 7.6 (pages 69 and 70)

1. $TRVG$ 2. TV

3. T 4. $TRHVG$

Lesson 8.1 (page 71)

1. $\frac{21}{4}$ 2. 3 3. $\frac{4}{3}$

4. 10 5. 6

Lesson 8.2 (page 72)

Verbal model:

$$\frac{\text{Sale price}}{\text{Number of CD's}} = \frac{\text{Price you pay}}{\text{Number of CD's purchased}}$$

Label: Price you pay $= x$ (dollars)

Equation:

$\frac{27.99}{3} = \frac{x}{5}$	Algebraic model
$\frac{27.99}{3} \cdot 5 = \frac{x}{5} \cdot 5$	Multiply both sides by 5
$46.65 = x$	Simplify

You will pay $46.65 for five CD's.

Lesson 8.3 (page 73)

1. Not similar–lengths of corresponding sides are proportional, but corresponding angles are not congruent.
2. Similar–corresponding angles are congruent and ratios of lengths of corresponding sides are equal. $\frac{6}{9} = \frac{8}{12} = \frac{10}{15} = \frac{2}{3}$
3. Not similar–corresponding angles are congruent, but lengths of corresponding sides are not proportional. $\frac{2}{2} = 1$ but $\frac{2}{4} = \frac{1}{2}$

Lesson 8.4 (pages 74 and 75)

1. Slope of $\overline{AB} = \frac{5-3}{6-4} = \frac{2}{2} = 1$

   Slope of $\overline{BC} = \frac{3-(-1)}{4-0} = \frac{4}{4} = 1$

   Slope of $\overline{AC} = \frac{5-(-1)}{6-0} = \frac{6}{6} = 1$

   $\triangle BDC \sim \triangle AEC$

2. Reasons
   2. Corresponding Angles Postulate
   3. Corresponding Angles Postulate
   4. Angle-Angle Similarity Postulate

3. Yes, by AA Similarity Postulate
4. No, $m\angle C = 70°$ but $m\angle F = 65°$
5. Yes, alternate interior angles $\angle 1 \cong \angle 2$ and $\angle 3 \cong \angle 4$. Also vertical angles $\angle 5 \cong \angle 6$. You can use AA Similarity Postulate.

Lesson 8.5 (pages 76 and 77)

1. $RT = 12$, $ST = 9$
2. $LM = 30$, $NM = 9$
3. SAS Similarity Theorem
4. AA Similarity Postulate
5. SSS Similarity Theorem
6. Reasons
   2. Base Angles Theorem
   4. Transitive Prop. of Congruence
   6. SAS Similarity Theorem

Lesson 8.6 (pages 78 and 79)

1. $\frac{WZ}{YZ}$ **2.** CA
3. AD **4.** DE **5.** c
6. d **7.** a **8.** b

Lesson 8.7 (page 80)

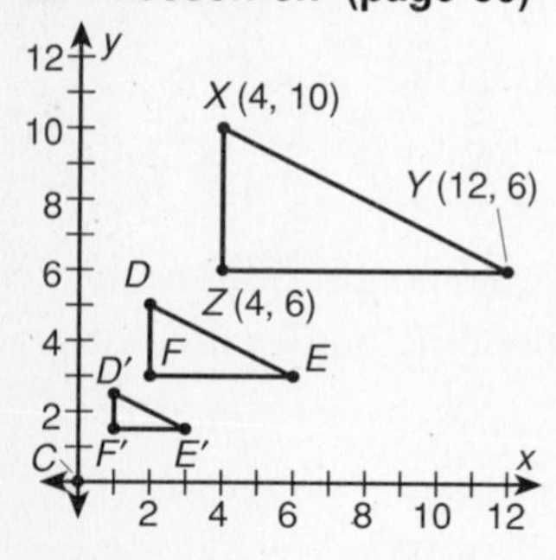

$$\frac{CX}{CD'} = \frac{CY}{CE'} = \frac{CZ}{CF'} = 4$$

Lesson 9.1 (pages 81 and 82)

1. $\triangle DBC$, $\triangle ABD$ **2.** BD, BD
3. AB **4.** DC, DC **5.** $\sqrt{12} = 2\sqrt{3}$
6. 2. Def. of $\perp$ lines
   3. Def. of right $\triangle$
   5. HL Congruence Theorem
7. *Reasons*
   1. Given
   2. Def. of a rectangle
   3. Def. of right $\triangle$
   4. Def. of right $\triangle$
   5. Diagonals of a rectangle are $\cong$.
   6. Reflexive Prop. of $\cong$
   7. HL Congruence Theorem

Lesson 9.2 (pages 83 and 84)

1. 24 **2.** $\sqrt{84} = 2\sqrt{21}$
3. $\sqrt{3}$ **4.** 26
5. Equation:

$840^2 + 350^2 = x^2$ — Algebraic model

$705{,}600 + 122{,}500 = x^2$ — Square 840 and 350

$828{,}100 = x^2$ — Simplify.

$\sqrt{828{,}100} = x$ — Take positive $\sqrt{\ }$

$910 = x$ — Simplify.

The distance from the balloon to the base of the observer is 910 feet. Since x is the hypotenuse of the right triangle, the answer is reasonable.

Lesson 9.3 (page 85)

1. b **2.** c **3.** a

Lesson 9.4 (page 86)

1. $x = \frac{8}{\sqrt{2}}$ or $4\sqrt{2}$, $y = \frac{8}{\sqrt{2}}$ or $4\sqrt{2}$
2. $a = 7\sqrt{3}$, $b = 14$
3. $g = 8$, $h = 8\sqrt{2}$
4. $j = \frac{6}{\sqrt{3}}$ or $2\sqrt{3}$, $k = 4\sqrt{3}$

Lesson 9.5 (pages 87 and 88)

1. $PR = 17$

   $\sin P = \frac{15}{17}$ $\sin R = \frac{8}{17}$

   $\cos P = \frac{8}{17}$ $\cos R = \frac{15}{17}$

   $\tan P = \frac{15}{8}$ $\tan R = \frac{8}{15}$
2. $XY = 9$

   $\sin Y = \frac{12}{15} = \frac{4}{5}$ $\sin Z = \frac{9}{15} = \frac{3}{5}$

   $\cos Y = \frac{9}{15} = \frac{3}{5}$ $\cos Z = \frac{12}{15} = \frac{4}{5}$

   $\tan Y = \frac{12}{9} = \frac{4}{3}$ $\tan Z = \frac{9}{12} = \frac{3}{4}$

3. $FE = 12$

$\sin D = \frac{12}{13}$ $\sin F = \frac{5}{13}$

$\cos D = \frac{5}{13}$ $\cos F = \frac{12}{13}$

$\tan D = \frac{12}{5}$ $\tan F = \frac{5}{12}$

4. $a \approx 6.4,\ b \approx 7.1$
5. $p \approx 11.5,\ q \approx 11.3$
6. $u \approx 3.7,\ v \approx 3.3$
7. $\tan Z = \frac{XY}{ZY}$, so $XY = (ZY)\tan Z = 220 \tan 40° \approx 184.6$ ft.

Lesson 9.6 (pages 89 and 90)

1. $m\angle P \approx 36.9°,\ m\angle R \approx 53.1°,\ RQ = p = 3$
2. $XY \approx 3.16,\ m\angle X \approx 71.57°,\ m\angle Y \approx 18.43°$
3. $ZY \approx 1.73,\ m\angle X = 60°,\ m\angle Y = 30°$
4. $XZ \approx 2.24,\ m\angle X \approx 41.81°,\ m\angle Y \approx 48.19°$
5. $RQ \approx 6.25,\ m\angle R \approx 38.65°$
   $RP \approx 8.01,\ m\angle P \approx 51.35°$
6. $RP = 15,\ m\angle R \approx 19.47°$
   $RQ \approx 14.14,\ m\angle P \approx 70.53°$
7. $RP \approx 4.24,\ m\angle R = 45°$
   $RQ \approx 3,\ m\angle P = 45°$

Lesson 10.1 (pages 91 and 92)

1. Answers will vary.

a.

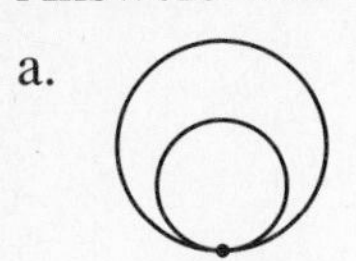

b.

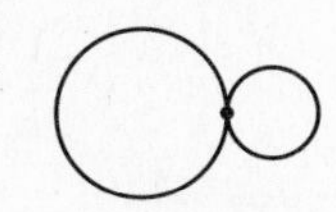

c.

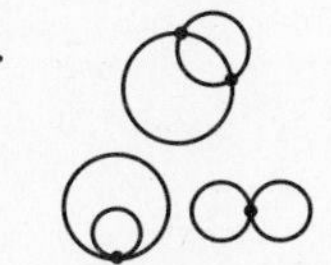

d.

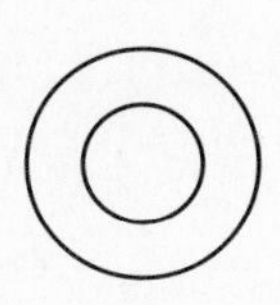

e.

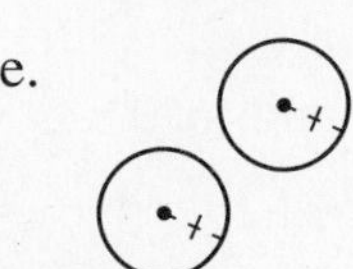

2. (a) 10.8 cm (b) 3.9 in.
   (c) $\frac{5}{6}$ in. (d) $2\sqrt{2}$ cm
3. Diameter 4. Secant
5. Common internal tangent
6. Center 7. Exterior point
8. Radius 9. Chord
10. Common external tangent
11. Point of tangency
12. Interior point
13. $\approx 245{,}652$ miles
14. $\approx 241{,}652$ miles

Lesson 10.2 (pages 93 and 94)

1.

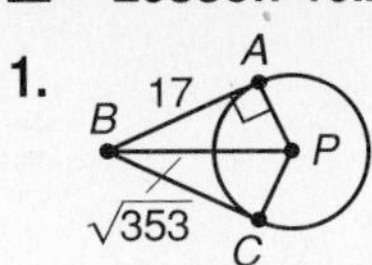

$$(PA)^2 = (\sqrt{353})^2 - 17^2$$
$$(PA)^2 = 353 - 289$$
$$(PA)^2 = 64$$
$$PA = 8 = PC$$

2. $PD = 10$ 3. $BC = 11$
4. $m\angle PQR = 90°$
5. Because $7^2 + 24^2 = 25^2$, you can apply the Converse of the Pythagorean Theorem to conclude that $\triangle PQR$ is a right triangle with hypotenuse $\overline{PQ}$. Then $\overline{PR} \perp \overline{RQ}$ and $\overleftrightarrow{QR}$ is tangent to circle P.

Lesson 10.3 (pages 95 and 96)

1. 80° 2. 165° 3. 345°
4. 100° 5. 115° 6. 245°
7. a. $\overset{\frown}{XZY}$ or $\overset{\frown}{XWY}$
   b. $\angle XPW$
   c. 80°
   d. $\overset{\frown}{XW}$ or $\overset{\frown}{WY}$
8. a. $\overset{\frown}{AB} \cong \overset{\frown}{CD}$ or $\overset{\frown}{AC} \cong \overset{\frown}{BD}$
   b. 85°
   c. $\overset{\frown}{AED}$
   d. 140°

Lesson 10.4 (pages 97 and 98)

1. $\overline{RT}$ 2. $\overset{\frown}{XS}$ 3. $\overline{ST}$
4. $\overset{\frown}{UV}$ 5. $\overline{SX}$ 6. b
7. d 8. a 9. e 10. c

Lesson 10.5 (pages 99 and 100)

1. 50° 2. 80° 3. 130° 4. 65°
5. 90° 6. 180° 7. 50° 8. 50°
9. 90° 10. 40° 11. 70° 12. 20°
13. 130° 14. 65°
15. $\angle A$ and $\angle C$, $\angle B$ and $\angle D$
16. No. Not all parallelograms have opposite angles that are supplementary. For example, $\angle A$ and $\angle C$ are not supplementary in this parallelogram.

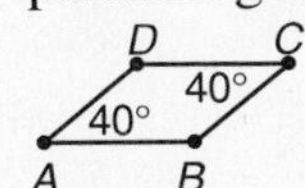

Lesson 10.6 (pages 101 and 102)

1. 82° **2.** $\angle MPQ$ or $\angle PTQ$
3. 66° **4.** $\angle LPQ$
5. 98° **6.** $\angle LPT$ or $\angle PQT$
7. e **8.** g **9.** h **10.** d
11. a **12.** b **13.** f **14.** c

Lesson 10.7 (page 103)

1. Center $(-3, 2)$, $r = 3$
2. Center $(5, 0)$, $r = \frac{1}{2}$
3. Center $(-1, 1)$, $r = \sqrt{3}$
4. $(x - 1)^2 + (y + 1)^2 = 25$

Lesson 11.1 (page 104)

1. 16 square feet **2.** 28 cm

Lesson 11.2 (page 105)

1. Area $\square PQRS = bh = 4(6) = 24$ square units
Area $\triangle PST = \frac{1}{2}bh = \frac{1}{2}(5)(6) = 15$ square units

2. a.

$$\text{Area } \triangle ACD + \text{Area } \triangle ABC = \text{Area } \triangle ABD$$
$$\tfrac{1}{2}(6)(7) + \text{Area } \triangle ABC = \tfrac{1}{2}(10)(7)$$
$$21 + \text{Area } \triangle ABC = 35$$
$$\text{Area } \triangle ABC = 14 \text{ square units}$$

b. Area $\triangle ABC = \frac{1}{2}bh$
$= \frac{1}{2}(4)(7)$
$= 14$ square units

3. $$\text{Probability} = \frac{\text{Area of } \triangle MPQ}{\text{Area of } \square MNOQ} = \frac{6}{32} = \frac{3}{16}$$

Lesson 11.3 (page 106)

1. $A = \frac{1}{2}(3)(18)$
$= 27$ square units

2. $A = \frac{1}{2}d_1d_2$
$= \frac{1}{2}(8)(12)$
$= 48$ square units

3. $A = \frac{1}{2}d_1d_2$
$= \frac{1}{2}(6)(15)$
$= 45$ square units

4. $A = \frac{1}{2}d_1d_2$
$= \frac{1}{2}(12)(14)$
$= 84$ square units

Lesson 11.4 (page 107)

1. $P = 4(6\sqrt{2}) = 24\sqrt{2}$ units
$A = \frac{1}{2}aP = \frac{1}{2}(3\sqrt{2})(24\sqrt{2}) = 72$ square units

2. $P = 6(2\sqrt{3}) = 12\sqrt{3}$ units
$A = \frac{1}{2}aP = \frac{1}{2}(3)(12\sqrt{3}) = 18\sqrt{3}$ square units

3. $P = 3(6) = 18$ units
$A = \frac{1}{2}aP = \frac{1}{2}(\sqrt{3})(18) = 9\sqrt{3}$ square units

4. Solve the right triangle shown to find the side of the square. Use the Area of a Square Postulate, $A = s^2$, to find the area.

Lesson 11.5 (page 108)

1. $C = \pi d = 12\pi \approx 37.7$ units

2. $16 = 2\pi r\left(\frac{120°}{360°}\right)$

$r = \frac{24}{\pi} \approx 7.64$ units

3. Length of $\widehat{AB} = 2\pi(5)\left(\frac{80}{360}\right)$
$= \frac{20\pi}{9} \approx 6.98$ units

Lesson 11.6 (page 109)

1. $A = \pi(5 \text{ cm})^2 = 25\pi \text{ cm}^2 \approx 78.54 \text{ cm}^2$

2. $A = \frac{300°}{360°}\pi(4 \text{ in.})^2 = \frac{40}{3}\pi \text{ in.}^2 \approx 41.89 \text{ in.}^2$

3. $A = \frac{60°}{360°}\pi(2 \text{ ft})^2 - \frac{1}{2}(2)\sqrt{3} \text{ ft}^2$
$= \left(\frac{2}{3}\pi - \sqrt{3}\right) \text{ft}^2 \approx 0.36 \text{ ft}^2$

Lesson 11.7 (page 110)

1. $\frac{P_1}{P_2} = \frac{3}{6} = \frac{4}{8} = \frac{1}{2}$

$\frac{A_1}{A_2} = \frac{1^2}{2^2} = \frac{1}{4}$

2. If the radius is tripled, the similar circles have corresponding parts in the ratio of 1 : 3, therefore, the ratio of the areas is 1^2:3^2 or 1 : 9. The circumference triples.

Lesson 12.1 (pages 111 and 112)

1. b **2.** a **3.** c

4. 6 vertices
Sketches vary.

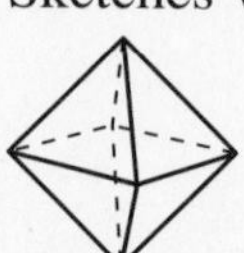

5. 6 edges
Sketches vary.

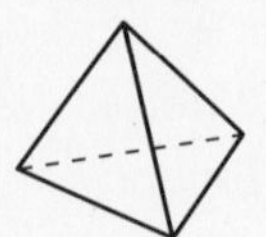

6. 6 faces
Sketches vary.

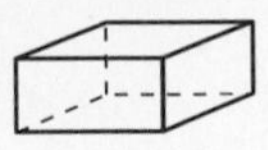

7. regular tetrahedron
8. regular dodecahedron
9. cube
10. regular octahedron

Lesson 12.2 (page 113)

1. 274 square units
2. 150 square units

Lesson 12.2 (page 114)

1. 150π sq. units ≈ 471.24 sq. units
2. 112π sq. units ≈ 351.86 sq. units
3. C in the formula $S = 2B + Ch$ is the circumference of the base. P in the formula $S = 2B + Ph$ is the perimeter of the base. The base of cylinders are circles which involve circumference. The base of prisms are polygons which involve perimeter.

Lesson 12.3 (page 115)

1. 85 square meters
2. $(9\sqrt{3} + 72)$ square in. ≈ 87.59 square in.

Lesson 12.3 (page 116)

1. 90π square units ≈ 282.74 square units
2. 60π square units ≈ 188.50 square units

Lesson 12.4 (page 117)

1. 343 cm^3 2. 96 ft^3
3. 768π in.$^3 \approx 2412.74$ in.^{3}
or $\frac{4}{9}\pi$ ft$^3 \approx 1.40$ ft^3

Lesson 12.5 (page 118)

1. $V = 48\pi$ cubic units ≈ 150.80 cubic units
2. $V = 240$ cubic units

Lesson 12.6 (page 119)

1. 100π square in. ≈ 314.16 square in.
2. $\dfrac{1372\pi}{3}$ cubic meters ≈ 1436.76 cubic meters
3. $V = 36\pi$ cubic feet ≈ 113.10 cubic feet
4. $V = \dfrac{128\pi}{3}$ cubic feet ≈ 134.04 cubic feet

Lesson 12.7 (page 120)

Given cube: $SA = 216$ in.^{2}
$V = 216$ in.^{3}
Similar cube: $SA = 1944$ in.^{2}
$V = 5832$ in.^{3}

Lesson 13.1 (page 121)

1.

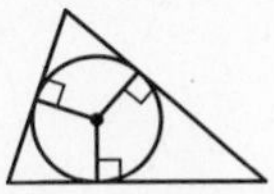

The locus of all points in a plane that are equidistant from the sides of a given triangle is the center of the inscribed circle.

2.

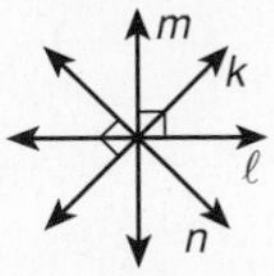

The locus of all points in a plane that are equidistant from perpendicular lines ℓ and m is two perpendicular lines n and k which bisect the right angles formed by lines ℓ and m.

3.

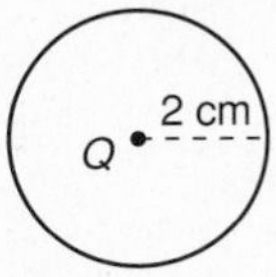

The locus of all points in a plane that are 2 centimeters from point Q is a circle with center Q and a radius of 2 centimeters.

4.

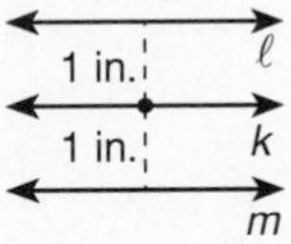

The locus of all points in a plane that are 1 inch from line k is two parallel lines ℓ and m that are 1 inch from line k.

Lesson 13.2 (page 122)

1. The locus is a line perpendicular to the plane of the triangle, containing the center of the inscribed circle.
2. The locus is a sphere with a radius of 2 centimeters and center Q.

Lesson 13.3 (page 123)

1. Points $(-3, 4)$ and $(3, 4)$

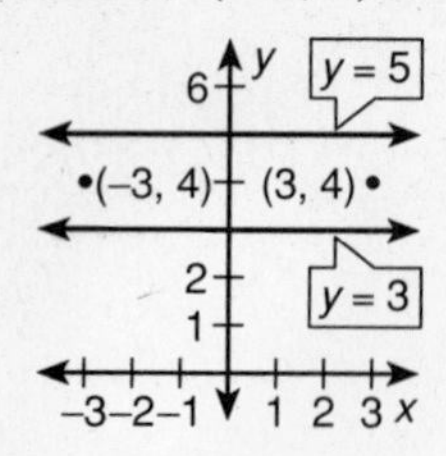

2. Points $(0, 5)$ and $(0, -5)$

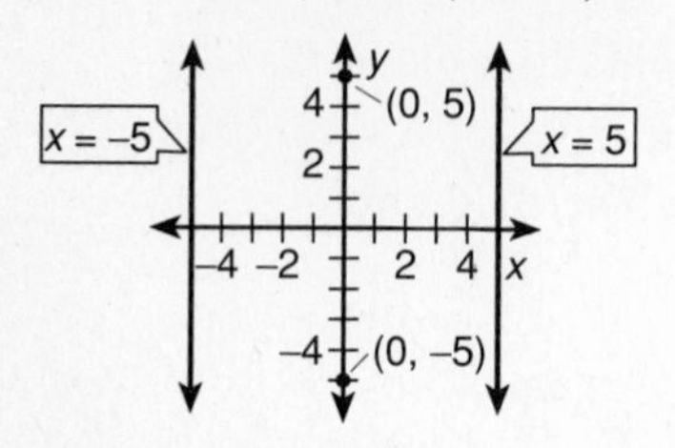

3. Point $(4, 0)$

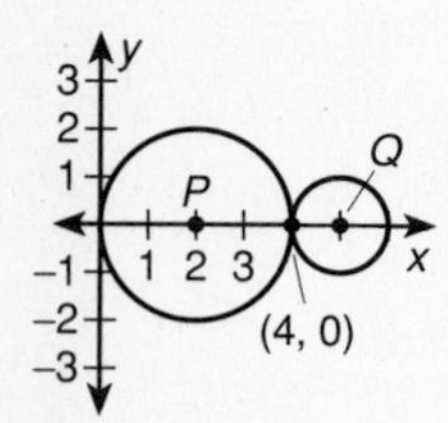

4. Points $(1, 1)$ and $(-1, -1)$

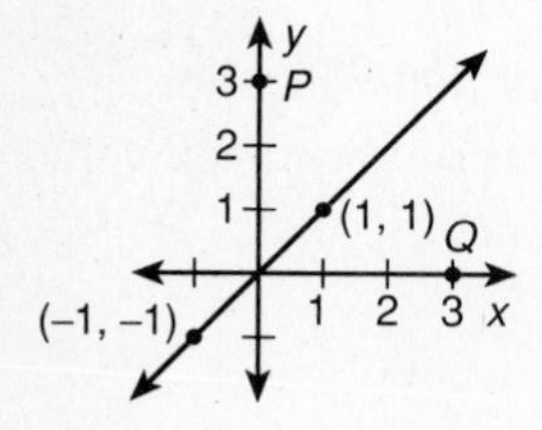

Lesson 13.4 (page 124)

1. $y = -x$

2.

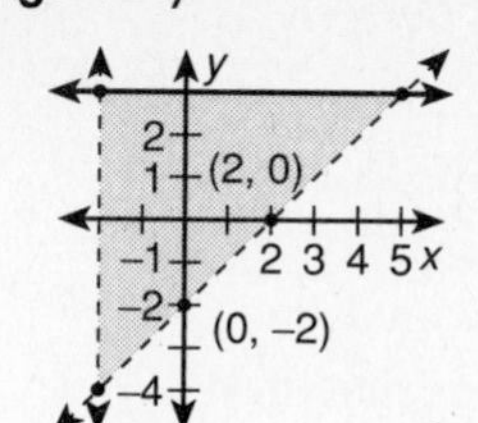

Lesson 13.5 (page 125)

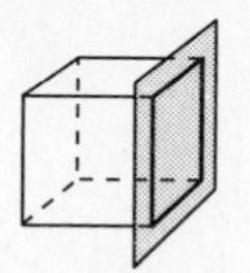

A square

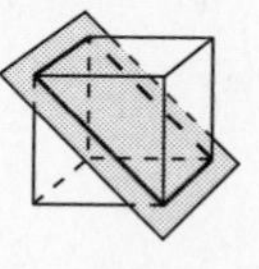

A rectangle

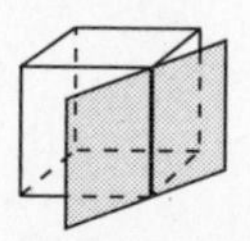

A line segment

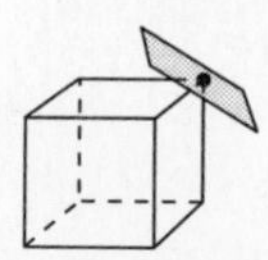

A point

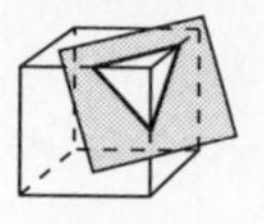

A triangle

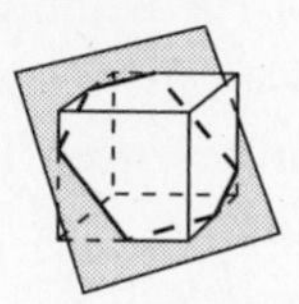

A hexagon

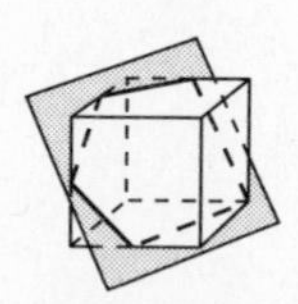

A pentagon

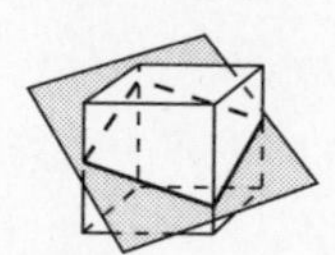

A parallelogram

Lesson 13.6 (page 126)

Answers will vary.